# THE EVOLUTION OF MAMMALIAN CHARACTERS

# The Evolution of Mammalian Characters

D. M. Kermack and K. A. Kermack
*Illustrated by A. J. Lee*

CROOM HELM
London & Sydney

KAPITAN SZABO PUBLISHERS
Washington DC

Croom Helm Ltd, Provident House, Burrell Row,
Beckenham, Kent BR3 1AT
Croom Helm Australia Pty Ltd, First Floor,
139 King Street, Sydney, NSW 2001, Australia

British Library Cataloguing in Publication Data

Kermack, D. M.
The evolution of mammalian characters.
1. Mammals — Evolution
I. Title II. Kermack, K. A.
599.03 QL708.5

ISBN 0-7099-1534-9

Published by KAPITAN SZABO PUBLISHERS
in the United States and Canada
1740 Lanier Place N W
Washington D C 20009

Library of Congress Cataloging in Publication Data

Kermack, D. M.
The evolution of mammalian characters.

Includes bibliographies and index.
1. Mammals, Fossil. 2. Mammals—Evolution.
I. Kermack, K. A. II. Title.
QE881.K45 1984 569 84-5783
ISBN 0-916845-01-X (Kapitan Szabo)

Printed and bound in Great Britain

# CONTENTS

# INTRODUCTION

This book is not intended to give a full and comprehensive account of the Mesozoic mammals, and nor is it intended as a handbook for research workers studying pre-Tertiary mammals. Our intention is to give an account of the origin and evolution of certain of the characters of the Mammalia. We have tried to portray the fossils we describe as the living animals they once were, not as dead bones. Our account ends with the end of the Lower Cretaceous, since by that time the major characters of the mammals had become established.

There exist a number of characters which, at the present day, are confined to the Mammalia. These include:

(1) a jaw articulation formed by the squamosal and the dentary;
(2) a chain of three bones, malleus, incus and stapes connecting the tympanic membrane to the inner ear;
(3) the presence of hair or fur;
(4) the presence of milk-glands in the female;
(5) the left aortic arch is the systemic arch;
(6) the phalangeal formula in both manus and pes is 2.3.3.3.3;
(7) some of the teeth have more than one root.

Of these characters (1) or (2) are sufficient by themselves to define a mammal; characters (6) and (7) are known to have been already in existence in some of the mammal-like reptiles — the ancestors of the mammals. Characters (3)*, (4) and (5) leave no impression on the skeleton, which by and large is what comprises the fossil record, so that we do not know if they were also present in the mammal-like reptiles. Character (5), however, on strong indirect grounds of inference, almost certainly existed in all the mammal-like reptiles as well as in the mammals.

Other characters, which are not confined to the mammals but which we always associate with them are:

(8) warm-bloodedness or homiothermy (also found in birds);
(9) viviparity (not found in monotremes, found in some representatives of all other living classes of gnathostomes except birds);
(10) growth by epiphyses (found in some lizards).

*Pterosaurs were claimed to be covered in fur. The 'fur' has now been identified as collagen fibres, in fairness to her, by the same worker who announced the original discovery.

The only one of these which could appear in the fossil record is (10), and, quite unexpectedly, it does not. It probably did not evolve until the mammals had come into existence, when the remarkably poor record of post-cranial skeletal material of Mesozoic mammals would explain its absence from the fossil record.

This book will deal with the interpretation of the facts as they are known from the fossil history of the mammals and the mammal-like reptiles; and it will only discuss soft parts when their nature can be unequivocally inferred from the skeleton (for example: the position of the tympanic membrane). In particular homiothermy and viviparity will not be discussed, nor will mammary glands.

There is a real danger that our ignorance of the evolution of the soft parts and physiological mechanisms will lead us to forget their importance to the living animal. A simple example of this, taken from a group alive at the present day, may be salutary. On their skeletal features cats are the most specialised of the living carnivores (Fissipedia). The cats were also the first group of the living fissipedes to differentiate: perfectly good cats were in existence at the beginning of the Oligocene, if not earlier. Dogs are less specialised skeletally than cats, and typical dogs do not become conspicuous until the Miocene. Cats are rather unspecialised physiologically, while dogs show great physiological specialisations for their cursorial mode of life; notably by being able to run fast for long periods without going into oxygen debt — a thing quite impossible to a cat. But this highly important physiological specialisation of the dogs leaves no record on the skeleton, and if the fissipedes were an entirely extinct group we would have no record that this specialisation ever existed.

Thus, by studying evolution by studying vertebrate palaeontology we inevitably get an incomplete and to some extent biased picture. We can only study the evidence we have, incomplete though it may be. But we should beware of the danger of so doing, even if we can do little about it.

We have concentrated on those aspects of the study of mammalian origins where we ourselves have made contributions to knowledge. This is a personal book. Jaw articulations, teeth and hearing consequently figure prominently; and in dealing with the fossil forms we have given most prominence to those upon which we have worked. In the references we have tried to include all papers on the subject published within the last ten years. We think that the book will be useful to those people who are interested in our ultimate ancestors, as well as to those carrying out research in the field.

The illustrations in the book are all by Mr A. J. Lee. The majority of them have been drawn either from the actual specimens or by redrawing and sometimes combining drawings in the original papers describing the material. We think that the illustrations will add greatly to the value of the book.

Finally we should like to express our thanks to our colleagues Mrs Frances Mussett and Miss P. M. Lees, without whose cooperation much of the research mentioned in this book would never have come to fulfilment.

K. A. Kermack,
Department of Zoology,
University College, London,
Gower Street,
London WC1E 6BT

D. M. Kermack,
Department of Pure & Applied Biology,
Imperial College,
Prince Consort Road,
London SW7 2BB

# CLASSIFICATION OF MESOZOIC MAMMALS USED IN THIS WORK

*Class Mammalia*

Subclass Atheria Kermack, Mussett and Rigney 1973

Order Triconodonta Osborn 1888
Suborder Eutriconodonta Kermack, Mussett and Rigney 1973
Family Triconodontidae Marsh 1887
Family Amphilestidae Kühne 1958
Suborder Morganucodonta Kermack, Mussett and Rigney 1973
Family Morganucodontidae Kühne 1958
Family Sinoconodontidae Mills 1971
Suborder Docodonta Kretzoi 1946
Family Docodontidae Simpson 1929

Order Multituberculata Cope 1884
Suborder Plagiaulacoidea Simpson 1925
Suborder Ptilodontoidea Sloan and Van Valen 1965
Suborder Taeniolaboidea Sloan and Van Valen 1965
Multituberculata *incertae sedis*
Family Haramiyidae Simpson 1947

Order Monotremata Bonaparte 1834

Subclass Theria Parker and Haswell 1897

Infraclass Pantotheria Simpson 1929

Order Eupantotheria Kermack and Mussett 1958
Suborder Amphitheria Kermack, Kermack and Mussett 1968
Family Kuehneotheriidae Kermack, Kermack and Mussett 1968
Family Amphitheriidae Owen 1846
Family Peramuridae Kretzoi 1960
Family Paurodontidae Marsh 1887
Suborder Dryolestoidea Butler 1939
Family Dryolestidae Marsh 1879
Suborder Symmetrodonta Simpson 1925
Family Amphiodontidae Simpson 1925
Family Spalacotheriidae Marsh 1887
?Family Aegialodontidae Kermack, Lees and Mussett 1965 (may be better classified as eutherian)

Infraclass Marsupialia Illiger 1811 }
Infraclass Placentalia Owen 1841 } Eutheria Gill 1872

# 1 THE ANCESTORS OF THE MAMMALS

## Earliest Vertebrates

To put the mammals in perspective and in the correct position in the fossil record it is necessary briefly to describe some of their known ancestors and emphasise points of particular interest and relevance about their structure before describing in detail the evolution of mammalian characters.

The first vertebrates to appear in the fossil record were the ostracoderms, armoured relatives of the present-day lampreys and hag-fishes. Their earliest remains are found in deposits of Ordovician or possibly Upper Cambrian age ($500 \times 10^6$ years ago)* (Table 1.1).

True fish with jaws, the acanthodians appeared at the end of the Silurian, but they do not resemble closely any present-day fish; by the Lower Devonian ($380 \times 10^6$ years ago), fish which are more recognisably the ancestors of groups now living, had come into existence. These fish fall into three groups (Figure 1:1): first the Palaeoniscoidea, which gave rise to the vast majority of modern bony fish living today; secondly the Dipnoi or lung-fish, which persist to the present time in rivers of Australia, South America and Africa and lastly, the Crossopterygii, which include the coelacanths, one of which, *Latimeria*, still lives in the Indian Ocean, and the ancestors of the Amphibia. Thus all the four-footed vertebrates or tetrapods arose from the Crossopterygii through the Amphibia (Figure 1.2).

The fossil remains of the early bony fish are found in the Old Red Sandstone of Devonian Age and laid down in fresh water. The conditions of deposition indicate that the whole landscape was of a fairly arid nature, with seasonal rivers which formed a series of pools of poorly aerated water in the dry season. The climate was probably like that of parts of East Africa at the present day. It was certainly not a desert. Under these conditions the respiration of dissolved oxygen from the water by means of the gills would become impossible, so, for the fish to survive, a means of utilising the abundant oxygen in the air had to be evolved. All three of the groups of bony fish — palaeoniscids, dipnoans and crossopterygians — evolved lungs for this purpose, as is shown by some of the descendants of all three groups at the present day still possessing them. Lungs, in actual fact, evolved as an adaption for living in poorly aerated fresh water, not as an adaption for living on land. In present day amphibians, lungs are best developed amongst those that live in poorly aerated water, not in terrestrial forms. Some terrestrial Amphibia lack

* Radiometric datings in this book follow *The Age of the Earth*: Institute of Geological Sciences, London, 1980.

| Approx. time B.P. ($\times 10^6$ years) | Era | Period | Approx. duration $\times 10^6$ | |
|---|---|---|---|---|
| 0 | Cenozoic | NEOGENE | | |
| | Cenozoic | PALAEOGENE | | |
| 100 | Mesozoic | CRETACEOUS | 77 | Birds; Mammals; Dinosaurs |
| | Mesozoic | JURASSIC | 53 | |
| 200 | Mesozoic | TRIASSIC | 35 | Therapsida |
| | Palaeozoic | PERMIAN | 50 | |
| 300 | Palaeozoic | CARBONIFEROUS | 65 | Pelycosauria |
| | Palaeozoic | DEVONIAN | 50 | First Amphibians |
| 400 | Palaeozoic | SILURIAN | 40 | First jawed fish |
| | Palaeozoic | ORDOVICIAN | 65 | First Vertebrates |
| 500 | Palaeozoic | CAMBRIAN | 70 | |
| 600 – 5,000 | Precambrian | | | |

**Table 1.1:** The Range in Time and First Appearance of the Main Vertebrate Groups

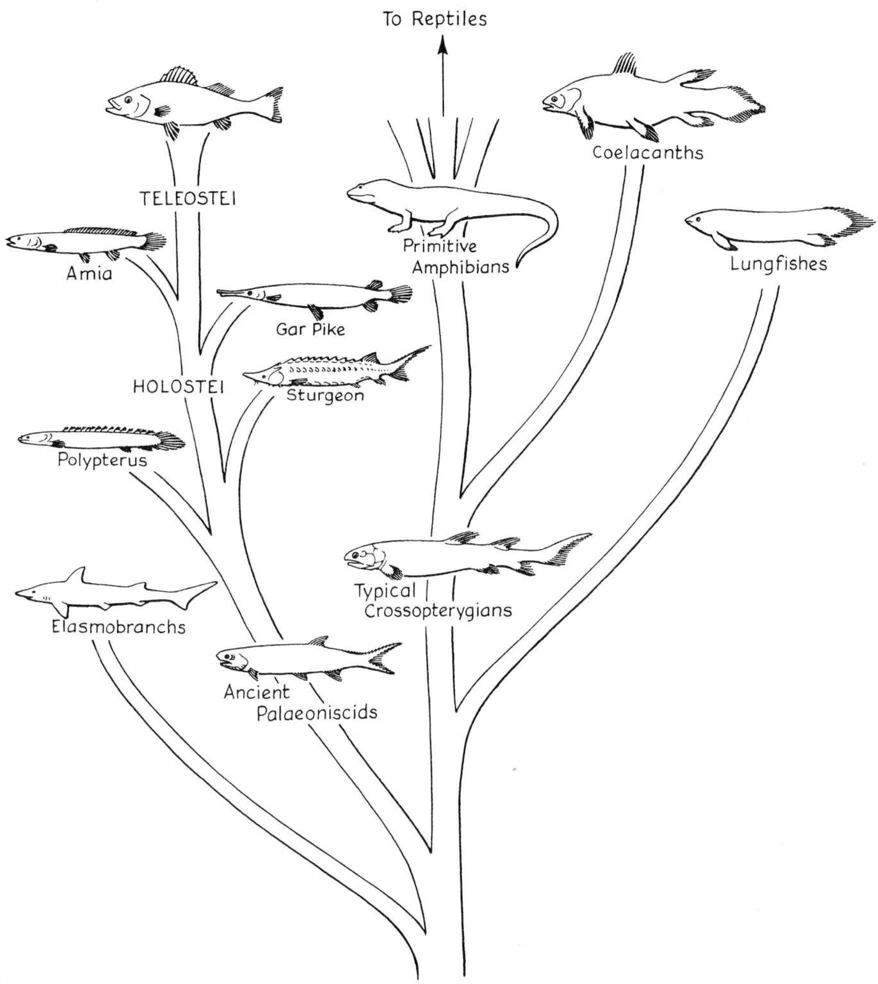


**Figure 1.1:** The Evolutionary Relationships of the Major Groups of Fish with the Amphibia

lungs altogether, and breathe through their moist skins; an example of this is the Plethodontidae, the largest family in numbers of species, of the urodeles.

During the late part of the Devonian period some relatives of the Crossopterygii evolved 5-fingered or pentadactyl limbs from their lobed-shaped paired pectoral and pelvic fins. These vertebrates, the ichthyostegids, were the first known Amphibia and their fossil remains are found in the Upper Devonian of East Greenland ($360 \times 10^6$ years ago).

The evolution of the Amphibia continued through the Carboniferous ($345 \times 10^6$ to $280 \times 10^6$ years ago). Although during this time the larger and better known Amphibia were all aquatic forms, there were also smaller ones, which moved onto the land becoming completely terrestrial. Their descendants

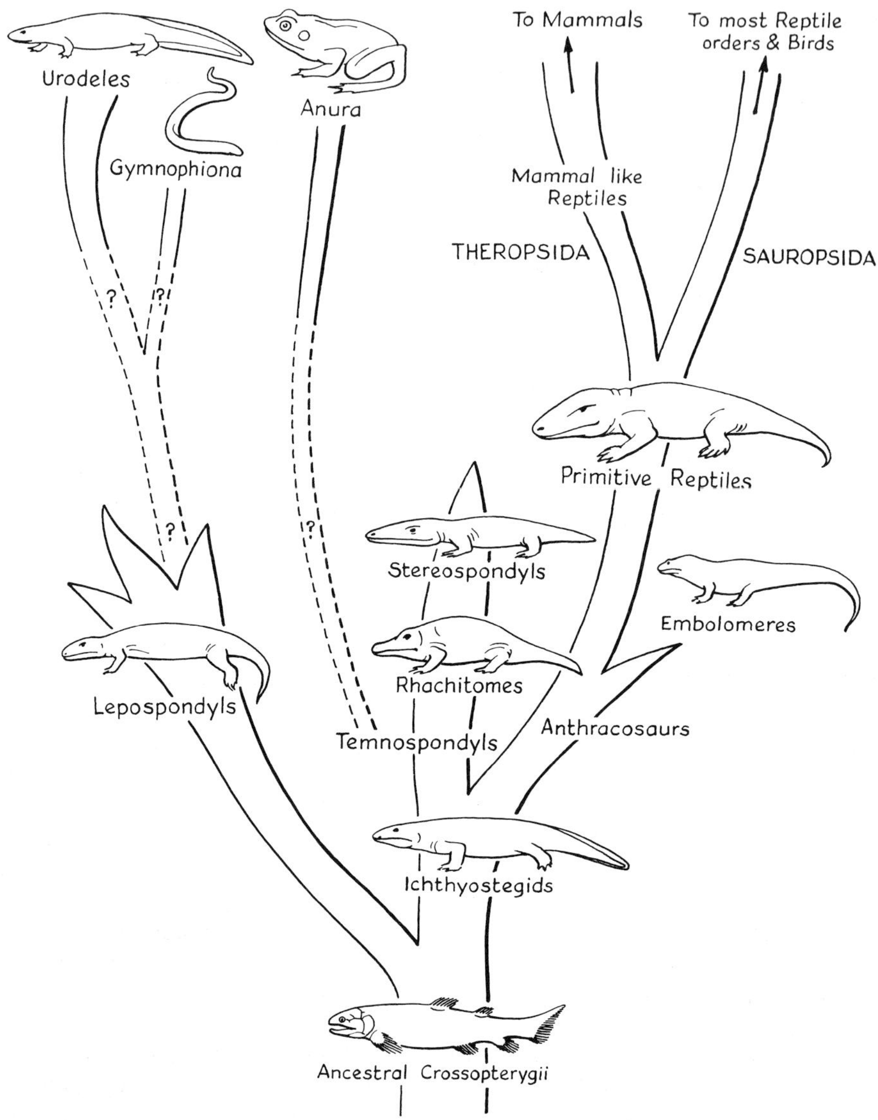


**Figure 1.2:** The Evolutionary Relationships in the Amphibia and Primitive Reptiles

became more and more emancipated from water so they can be regarded as the earliest reptiles. This is the next stage in the evolutionary history of the early mammals.

## Anamniotes and Amniotes

The fundamental difference between the Amphibia and the Reptilia lies in the development of the embryo. In those amphibians which have not become too specialised in their mode of reproduction, the eggs are laid in water, where they hatch when the embryo is at quite an early stage of development. The young amphibian emerges as a herbivorous and aquatic larva or tadpole, which has a free-living existence until it is ready to undergo metamorphosis to the adult form. This is the mode of development of many common Amphibia of the northern temperate regions and consequently will be familiar to most readers of this book.

Many Amphibia lay large yolky eggs so that the embryo hatches at a much later stage of development. Sometimes there is so much yolk that hatching is consequently so delayed that metamorphosis takes place within the egg, with the young amphibian emerging as a miniature replica of the adult. But even in these cases the outer membrane of the egg remains freely permeable to water, so that unless the egg remains in water or in a permanently humid atmosphere fatal desiccation must take place.

The reptiles have solved this problem in a simple and direct way. The egg is enclosed in a tough shell, freely permeable to oxygen and carbon dioxide, but not freely permeable to water vapour. The degree of permeability of the shell to water vapour varies greatly from species to species, but in some Chelonia which lay their eggs in dry sand, the shell is almost completely impermeable to water vapour. At an early stage of the development of the reptilian egg an extra-embryonic membrane grows around the embryo and finally encloses it in a fluid-filled sac. This provides the embryo with 'its own private pond' in which to develop. The sac is called the *amnion* and the fluid within it the *amniotic fluid* (Figure 1.3).

Thus the reptiles emancipated themselves from the dependence on water for reproduction which has always tied the Amphibia to aquatic or humid habitats, and so were able to complete the conquest of dry land. Those reptiles and mammals which are viviparous retain the amnion to supply a controlled safe environment for the development of the embryo. In consequence reptiles, and their descendants the birds and the mammals, are frequently grouped together as *amniotes* to contrast them with the Amphibia which never acquire an amnion.

The Reptilia are a large and diverse group of amniotes which first appear in the Upper Carboniferous (Table 1.1). Before discussing in detail the ancestors of the mammals, it will be useful to consider the broad relationships within the amniotes as a whole.

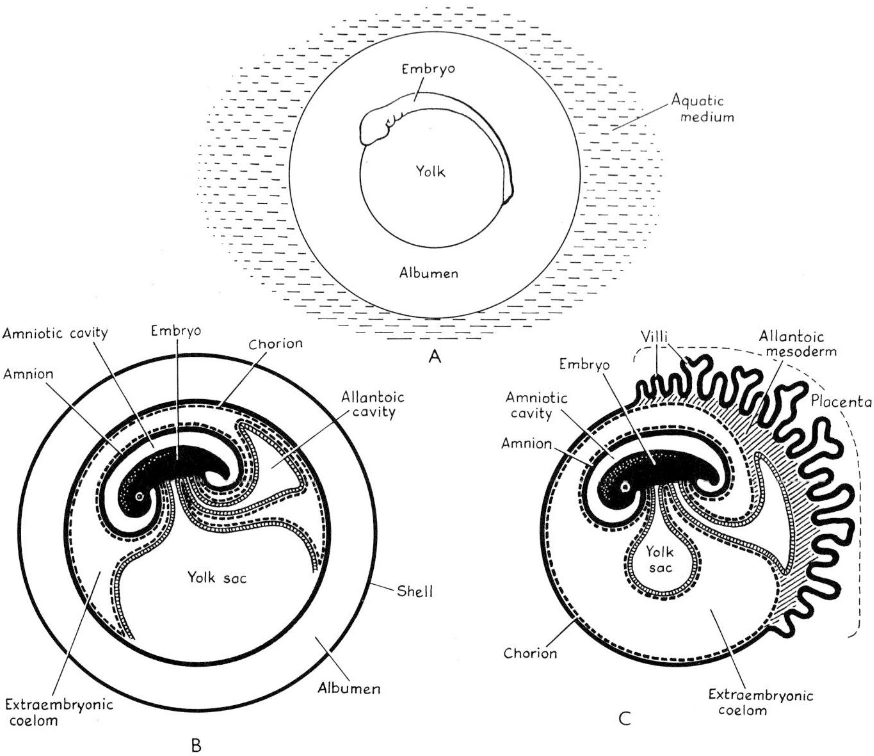


**Figure 1.3:** The Anamniote Egg and Embryo of a Frog (A), Compared with the Amniote Egg of a Bird (B) and a Mammal (C)

## Sauropsida and Theropsida

In 1916 Professor E.S. Goodrich suggested that all the amniotes could be divided into two groups, which must have arisen quite independently of each other, either from very primitive reptiles or directly from the Amphibia. He based his suggestion on a study of the great arteries which carry blood away from the heart. The vessel (systemic arch) which carries oxygenated blood from the left ventricle of the heart lies on the right side of the body in modern reptiles and birds and on the left side in mammals (Figure 1.4). It is clearly quite impossible for the condition found in birds and modern reptiles to have arisen from that found in mammals or *vice versa*. Both must have arisen from the condition found in the Amphibia where both right and left arches are equally developed. The group which includes all living, and most extinct, reptiles and the birds Goodrich called the Sauropsida; and the group including the mammals and their ancestors the mammal-like reptiles, which must have had the same pattern of aortic arches as the mammals, the Theropsida.

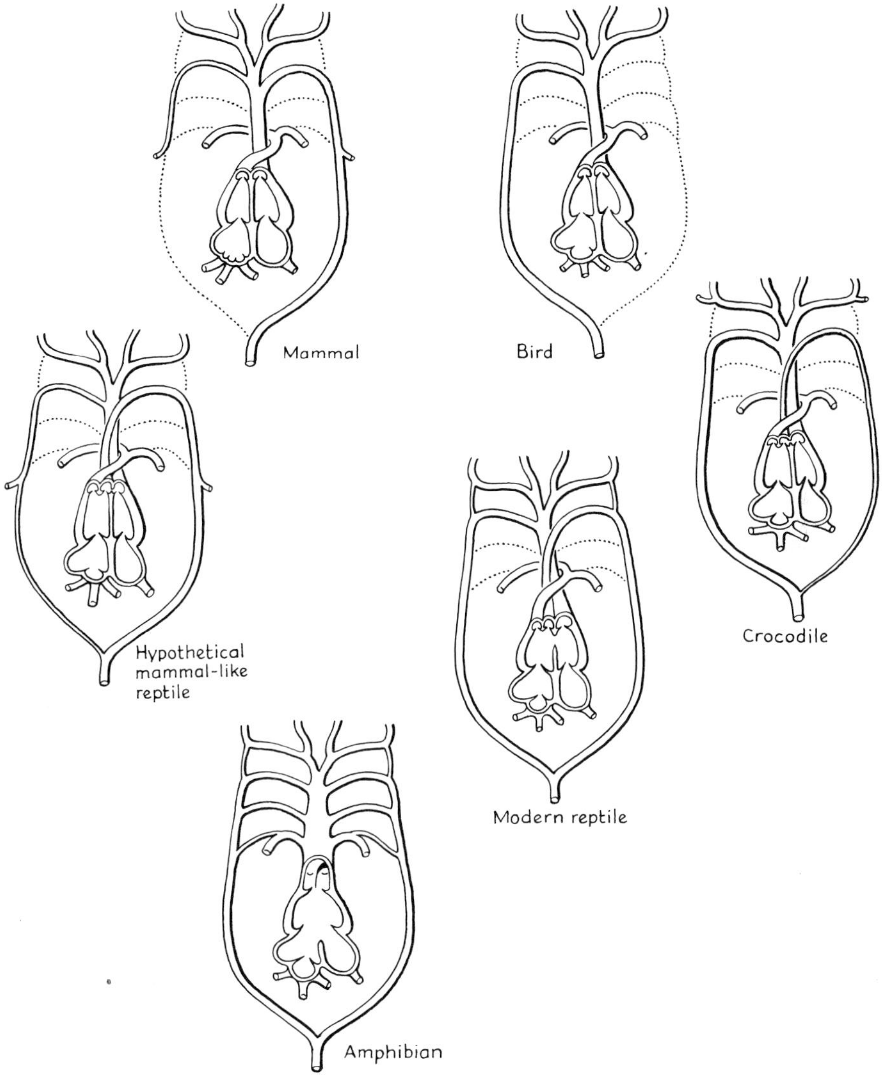


**Figure 1.4:** The Arrangement of the Main Systemic Arches in Tetrapods
Source: After Goodrich, 1916

In 1951 Professor D.M.S. Watson enlarged upon this concept of Goodrich. Watson pointed out that the sloping otic notch of the early Amphibia (labyrinthodonts) gave rise to mechanical weakness at the hinder end of the skull. This weakness could be eliminated if the quadrate bone, which forms the fixed part of the hinge of the lower jaw and is steeply inclined in the labyrinthodonts, became vertical. As Figure 1.5 shows, this can take place in one of two ways: either the upper end of the quadrate forms the pivot

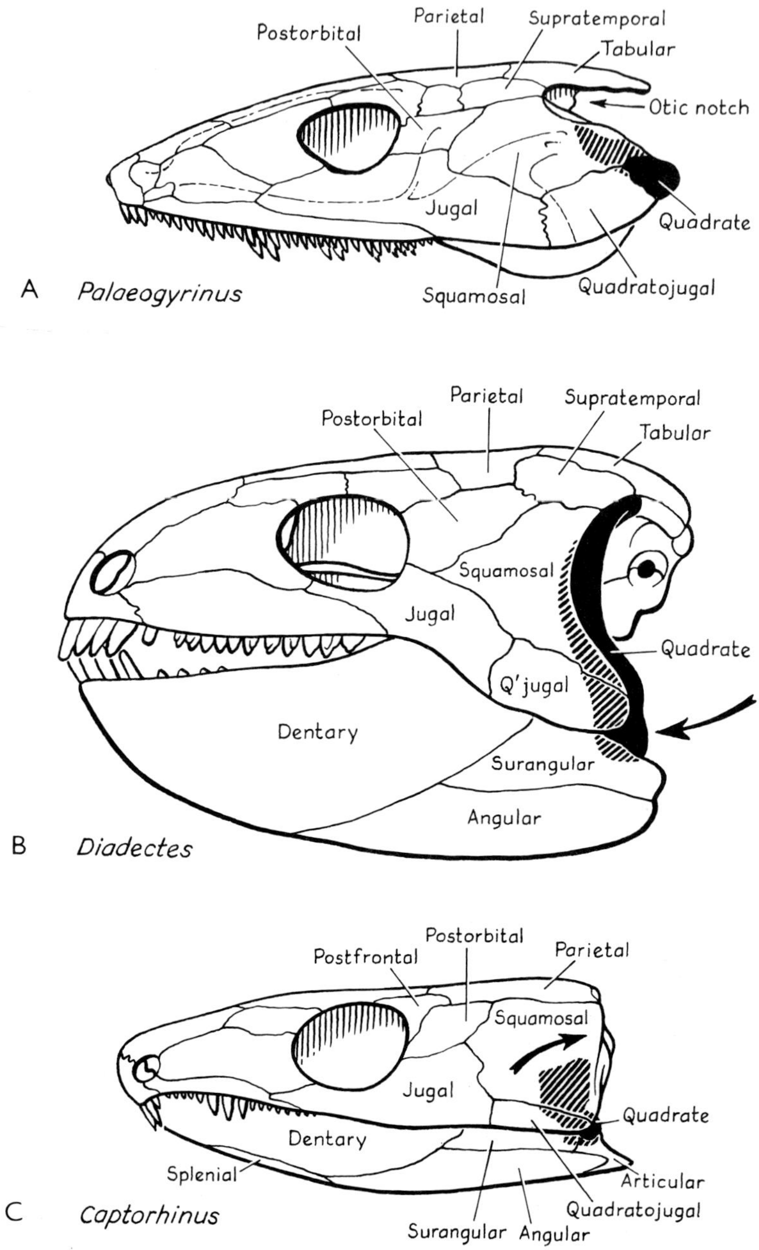


**Figure 1.5:** The Rotation of the Sloping Labyrinthodont Quadrate (A) in Primitive Reptiles. The shaded portion of the quadrate lies underneath the dermal bones of the skull roof. The arrows indicate that in the Sauropsida the quadrate becomes vertical by the rotation of its lower end forwards (B), leaving a shallow otic notch. In the Theropsida the upper end of the quadrate is rotated forwards (C) obliterating the otic notch

and the lower end swings forward (B); or the lower end forms the pivot and the upper end swings backward (C). In the former case the otic notch is opened out and the tympanic membrane is then attached along its anterior border to the posterior edge of the quadrate. In the latter case the otic notch is obliterated. Thus there is nowhere to put the tympanic membrane which must have been absent. As Watson showed, the members of Goodrich's group Sauropsida either fall into the condition shown in Figure 1.5B, or they are clearly derived from animals which had this condition. The Theropsida all fall into the category illustrated in Figure 1.5C, or are clearly derived from it.

All amniotes with the exception of the Cotylosauria and the Chelonia (tortoises and turtles) have lateral fossae in the temporal region of the skull roof. These have the function of accommodating the jaw-closing muscles when they contract. The positions of the fossae are extraordinarily constant within the major groups of amniotes and they are used to classify the Reptilia into their major taxons. These fossae can occur in the positions shown in Figure 1.6, which also gives the name given to the type of skull in each case. The Sauropsida show all these types except the synapsid. Here there is but a single opening in the skull low down in the cheek region. The fossa is bounded above by the squamosal and postorbital bones and below by the jugal and squamosal. Only the skulls of mammal-like reptiles and mammals have the synapsid type of skull.

It must be noted that we are talking here about the primitive type of skull in the group. Some of the more specialised mammal-like reptiles, and the mammals themselves, do not strictly have the synapsid type of temporal fossa, although their skulls are quite clearly derived from ancestors which did. Despite this, their skulls are all formally called synapsid.

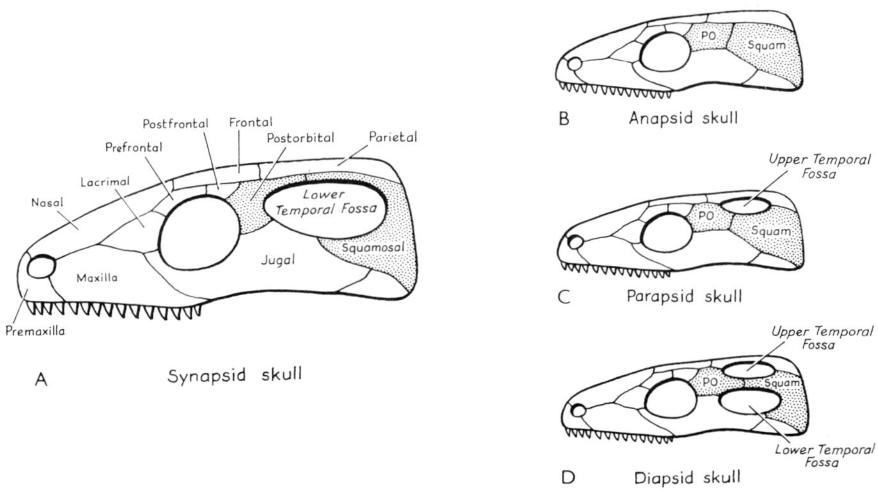


**Figure 1.6:** The Primitive Arrangement of the Temporal Fossae in the Different Patterns of Reptilian Skull. Mammals have skulls which can be derived from the synapsid condition

## The Synapsida

The mammal-like reptiles and the mammals share three common characters making them a closely knit group separated off from the rest of the amniotes. The first character is the presence of the synapsid type of skull or a modification of it; the second, the pectoral endochondral girdle with three ossifications and the last, the development of the left systemic arch to carry blood from the heart to all parts of the body. The second character the mammals and mammal-like reptiles share with the obviously closely related captorhinomorph reptiles but with no other animals. The last of the three characters is the least satisfactory as it is found only in living mammals, but a study of comparative vertebrate anatomy and embryology indicates that it is correct to assume that the mammal-like reptiles would have showed at least greater development of the left systemic arch than the right one. Nevertheless as a general principle, osteological characters, which show on fossil material, are a far more satisfactory basis for the classification of the major vertebrate groups on account of their long palaeontological history.

As all the mammal-like reptiles have the synapsid type skull, they are grouped together in the subclass Synapsida in the class Reptilia (Williston, 1925; Romer, 1956) and the Synapsida and the mammals together make up the Theropsida.

The Synapsida have a long and important history. They are among the earliest known reptiles and during most of their history were the dominant group of land animals, playing much the same role in the world fauna as the mammals, their descendants, do today. In other words there were times when the vast majority of the larger terrestrial forms were synapsids, both carnivores and herbivores, as is true of the present-day Mammalia, which have been dominant since the beginning of the Cenozoic era, some seventy million years ago. The mammal-like reptiles first appear in the Upper Carboniferous of North America. According to Romer and Price (1940) *Clepsydrops* from the Stephanian near Danville, Illinois, is the earliest synapsid whose position is certain, both as to its synapsid nature and its geological age. Carroll (1964) described a specimen which is probably a pelycosaur of Westphalian age from the Joggins Reef and therefore older than that from the Danville deposits. If the identification of this specimen be correct, then *Protoclepsydrops haplous* is not only the earliest known synapsid but also one of the earliest known reptiles. The Synapsida rapidly established themselves as the major group of land animals, a position they held throughout the Permian and the first half of the Triassic, a period of about $70 \times 10^6$ years. In the second half of the Triassic they were suddenly eclipsed by the newly emerged dinosaur orders. About this time the Mammalia came into existence, so the Theropsida continued as small vertebrates; the largest known Mesozoic mammal was about the size of a polecat. Although small, they may well have been individually numerous and played an important role in the life of their time. This state of affairs lasted for over a hundred million years, but with the end of the Cretaceous the dinosaurs declined even more rapidly than did the mammal-like reptiles in the Trias, and vanished completely from the Earth. The Theropsida took their place and became, for the second time and until the present day, the dominant group of land vertebrates. The causes of the decline of the Synapsida in the Trias and their even more spectacular recovery at the end of the Cretaceous remain a mystery.

These early reptiles found alongside the Upper Carboniferous amphibians show characteristics, indicating that they have already moved some way along their evolutionary path. The practical difference between an amphibian and a reptile is the dependence of the former to some extent upon water for reproduction, whilst the latter is almost completely emancipated from it. Thus reptiles can colonise the dry land in a way that was quite impossible to their amphibian ancestors and contemporaries. Consequently, early reptiles must have exploited this advantage and lived completely terrestrial lives away from the coal-measure swamps and the competition from amphibians inhabiting them. They probably lived inland upon the drier hills above the lowland swamps. Thus reptilian remains would probably only be found in the swamp deposits, if they had been carried in streams and rivers running down from the hills to the coastal and estuarine swamps. This is the most likely explanation of the presence of a few reptile remains in the Joggins Reef deposits described by Carroll and also the 'sophistication' of those reptiles found amongst the amphibian remains. It is unfortunate that dry-land deposits are so rare in the Upper Carboniferous, but they are found less rarely at the very end of the Upper Carboniferous and much more commonly in the Permian. It is for this reason that the part of reptilian evolutionary history occurring in the Permian is well documented. The whole situation is an excellent illustration of the point that there must be easily accessible deposits of the right age available before there is any hope of finding the appropriate fossils to fill gaps in the evolutionary history of an animal group. Later chapters will provide additional examples and it can be said that the lack of suitable deposits has bedevilled the study of Mesozoic mammals.

## The Pelycosauria

Romer (1956) divides the primitive forms of the sub-class Synapsida into two groups, firstly into the order Pelycosauria and secondly into the order Therapsida containing the later and more mammal-like forms. As Romer himself explains, the distinction between members of the two groups is to some extent exaggerated by the vagaries of the geological record, as they are separated from each other in space as well as in time. The earlier Pelycosauria are mainly known from the USA, whereas the later Therapsida are found in South Africa and Russia.

The pelycosaurs are already common in the early Permian and show considerable diversity of structure, which enabled them to live in different habitats (Table 1.2). This kind of diversity is referred to as a 'radiation'. The division of the pelycosaurs into three sub-orders, the Ophiacodontia, Edaphosauria and Sphenacodontia gives some idea of the degree of radiation that occurred in these early reptiles. The ophiacodonts seem to have been in their prime at about the Carboniferous-Permian boundary and were probably ancestral to members of the other two sub-orders, although a high proportion of them seem to have been specialised for eating fish and were secondarily aquatic. The edaphosaurs were terrestrial herbivores, whilst the sphenacodonts were carnivorous and probably preyed on the former. Both these sub-orders reached their acme in the Lower Permian, rather later than the ophiacodonts. The edaphosaurs became such a specialised side branch that they were soon doomed to extinction. The specialisations for a carnivorous way of life are less extreme than those demanded of plant-eaters. In consequence it is the carnivores and insectivores which give rise to new groups of vertebrates, which in their turn radiate into all habitats and modes of life. Herbivores are so specialised that they can only give rise to still more specialised herbivores, until the whole lineage becomes extinct. This is what happened to the edaphosaurs, but the sphenacodonts on the other hand later gave rise to more advanced mammal-like reptiles. In evolution, it is never 'the meek that inherit the earth'.

So the Sphenacodontia were the dominant carnivores of the Lower Permian and thanks to the work of Romer and his associates, *Dimetrodon* from the Red Beds of Texas is the best known of them. It was a large reptile about three metres long, standing much higher off the ground than the early pelycosaurs (Figure 1.7). The shape of the limb bones, and other features of the skeleton indicate that it was a fairly active animal, indeed, it gives the impression of having been a formidable and efficient carnivore. The most

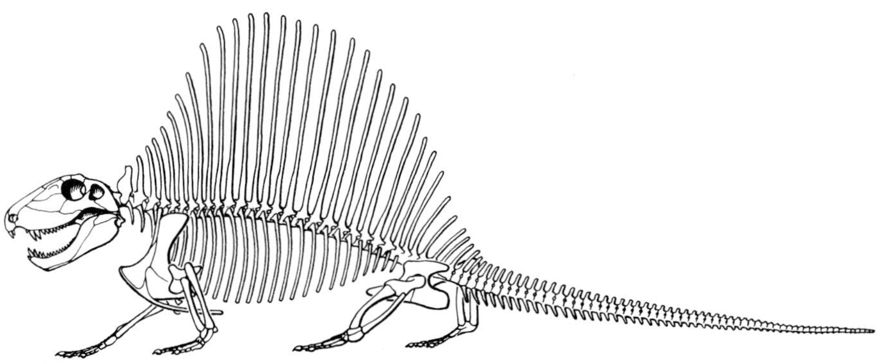

**Figure 1.7:** The Skeleton of *Dimetrodon*
Source: After Romer and Price, 1940

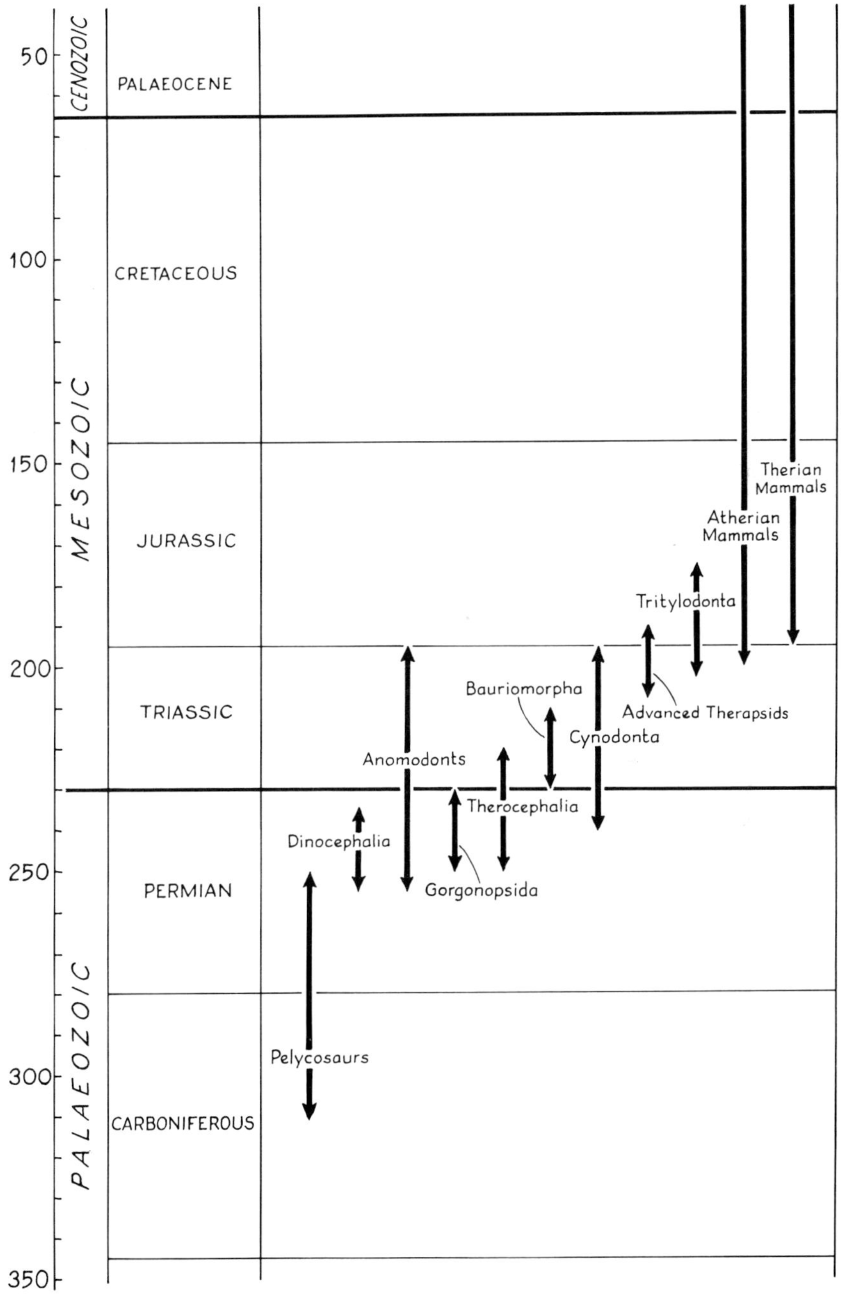


**Table 1.2:** The Range in Time of the Main Groups of Mammal-like Reptiles

striking feature about the skeleton was the enormous elongation of the neural spines of the dorsal vertebrae, so artists draw reconstructions of the animal showing a large dorsal 'sail' running the length of its back. This bizarre development of the neural spines is not unique to *Dimetrodon*, it was also shown in the coeval *Edaphosaurus* as well as in *Spinosaurus*, a dinosaur living 150 million or so years later (Upper Cretaceous of Egypt, Stromer, 1915) and *Spinosuchus*, a small quite unrelated dinosaur from the Upper Trias. Unexpectedly *Lupeosaurus*, an edaphosaur, had elongated spines more like those of *Dimetrodon* than those of its closer relative *Edaphosaurus*. There is a wealth of theories regarding the function of the 'sail': these are that it acted as a radiator for temperature control, as a sail in locomotion or it was for defence or camouflage, and so on. Yet there is only one thing that can be said with certainty, and that is that these elongated spines were useful to their possessor and must have evolved through the normal action of Natural Selection upon favourable gene mutations. These greatly elongated neural spines are found in such diverse animals that no theory of the chance fixation of a freak mutation can be creditable. Unfortunately it is almost impossible to assign a function to a character found in an extinct animal but not found in any living one. Similarly greatly elongated spines support the dorsal fin in the living sail-fishes (*Istiophorus*); and to date no satisfactory explanation of the function of this enlarged dorsal fin has been given.

Figures 1.8 and 1.9 show different views of the skull of *Dimetrodon*. The lateral temporal fossae are quite large and retain the primitive relationship with their surrounding bones, the post-orbital and squamosal forming the upper boundary of each and the jugal the lower. Their size is correlated with the powerful jaw muscles that are associated with a carnivorous way of life. The occipital plate is steeply inclined, a feature associated with the development of powerful neck muscles for keeping the head raised. The lateral view of the skull shows that it is deep, particularly in the snout region due to the enlargement of the maxilla to accommodate the roots of the large canines. The premaxillary teeth are not as enlarged as some of those in the maxilla, so, as there has been no corresponding increase in depth in the premaxilla, the line of the upper jaw shows a well marked 'step'. The teeth of the lower jaw have also increased in size and so, like the maxilla, the dentary has also increased in depth to accommodate their roots. This phenomenon occurs in the jaws of many Permian mammal-like reptiles.

The ventral view of the skull shows the palate and illustrates well the enlargement of a pair of teeth towards the anterior end of each maxilla. These enlarged teeth are the forerunners of the mammalian canines and this will be discussed further in Chapter 5. The large stapes running from the back of the braincase outwards to the well-developed quadrate can also be seen in this view. In an animal such as *Dimetrodon*, with a skull often more than 300 mm long, the stapes is a massive bone, as large as a human thumb, which ended in a roughened surface, indicating that it would have continued as a

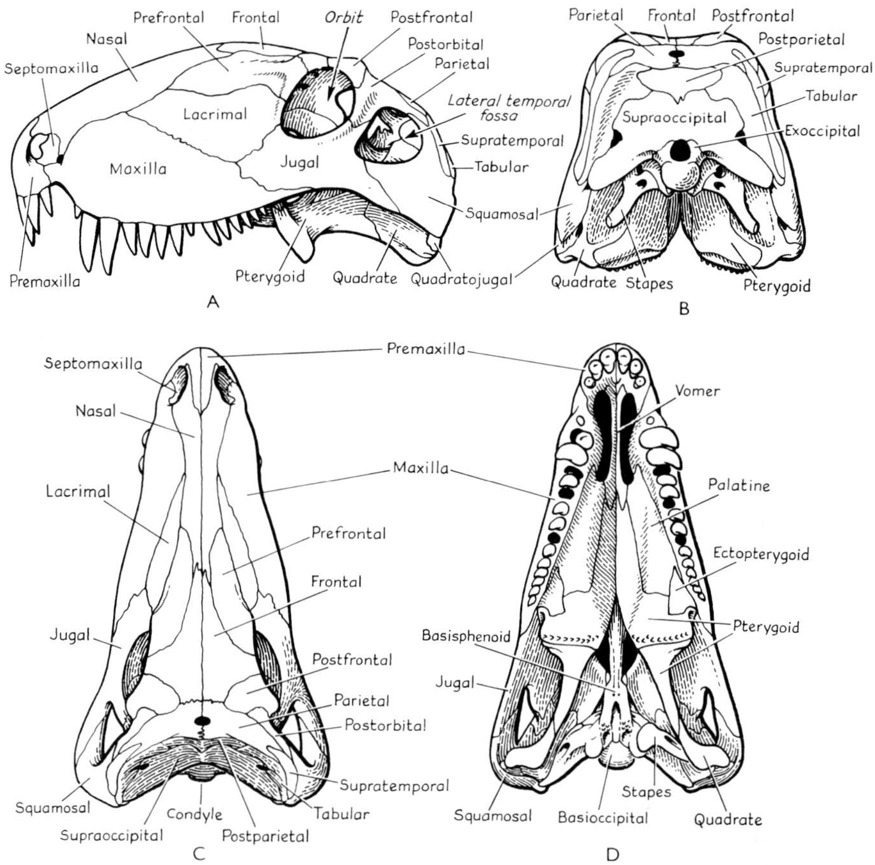


**Figure 1.8:** The Skull of *Dimetrodon*. A, lateral view; B, occipital view; C, dorsal view and D, ventral view
Source: After Romer and Price, 1940

cartilaginous extra-stapes in life. It is also visible in the occipital view in Figure 1.8. This figure shows that the posterior temporal fossae are reduced to two small foramina. The quadrate, which plays a key role in the evolutionary history of the mammals, is a great vertically placed bone. Figure 1.9A shows the skull with the roof removed on one side, so that it is possible to look inside. The palate is a great domed structure falling almost vertically posteriorly to the commencement of the pterygoid flanges. To some extent this doming is simply due to the pulling down of the lower border of the maxilla to accommodate the roots of the greatly enlarged anterior teeth, but it is probably also a direct adaptive change. It would have kept open an airway when the lower part of the buccal cavity was blocked by recently captured prey, probably still alive and struggling. The figure also shows the great transverse process of the pterygoid at the back of the skull on each side,

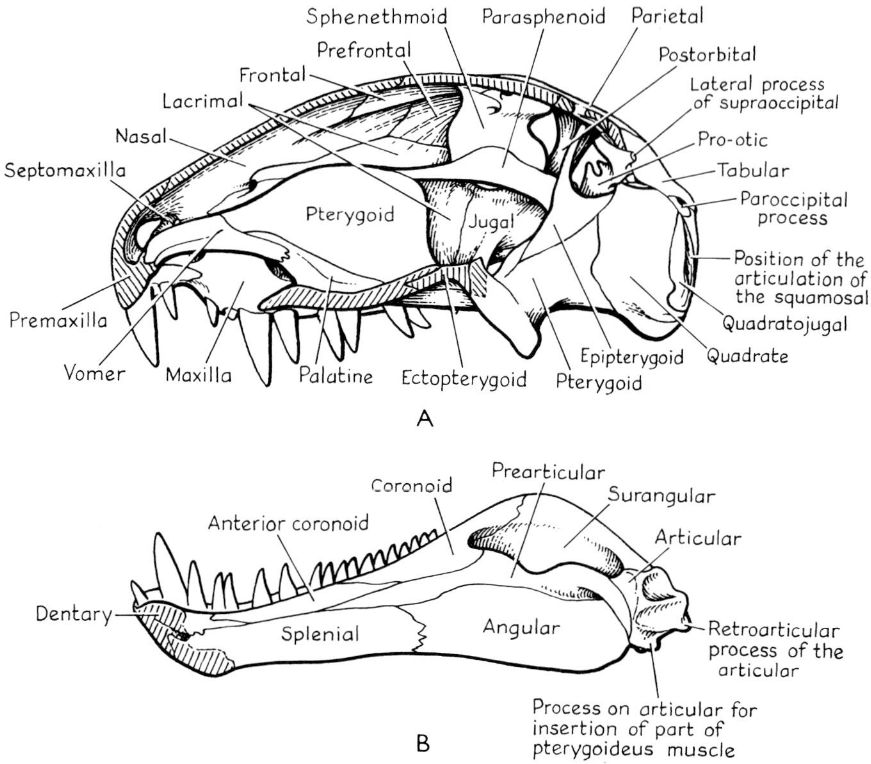


**Figure 1.9:** The Skull of *Dimetrodon*. A, lateral view with the skull roof on the left side removed, showing the doming of the palate. B, medial view of the lower jaw
Source: After Romer and Price, 1940

and above this the expanded base of the epipterygoid can be seen running back to the quadrate. Both these structures have important roles to play in the history of the origin of the Mammalia.

Figure 1.9B shows well the complex nature of the lower jaw. Its upper margin is formed mostly by the dentary, which bears all the lower teeth. Posteriorly the surangular continues the line of this upper margin. The lower margin of the jaw is formed by the splenial and the angular. At the posterior end of the jaw is the articular, which forms, with the quadrate, the hinge of the jaw. The articular has a concave articulating surface, while the corresponding surface of the quadrate is convex. Between the articular and the angular is the prearticular, lying along the medial side of the dentary. Above the prearticular are the two coronoids. These bear on the transverse process of the pterygoid, and prevent the lateral displacement of the lower jaw on opening. A pair of coronoids is a primitive condition, and most later reptiles and primitive mammals have but a single coronoid.

In the pelycosaurs we see the beginning of the differentiation of the teeth into incisors, canines and cheek-teeth. This is described in detail in Chapter 5.

## References

Carroll, R. L. (1964) 'The Earliest Reptiles', *J. Linn. Soc. (Zool). Lond., 45*, 61–83

Goodrich, E. S. (1916) 'On the Classification of the Reptilia', *Proc. Royal Soc. Lond.* series B, *89,* 261–76

Romer, A. S. (1956) *Osteology of the Reptiles*, The University of Chicago Press, Chicago

Romer, A. S. & Price, L. W. (1940) 'Review of the Pelycosauria', *Geol. Soc. Amer. Sp. Papers*, no. 28

Stromer, E. (1915) 'Das Original des Theropoden *Spinosaurus aegyptiacus', Abh. Bayer. Acad. Wiss., 18*, 1–32

Watson, D. M. S. (1951) *Paleontology and Modern Biology*, Yale University Press, New Haven

Williston, S. W. (1925) *The Osteology of the Reptiles*, Harvard University Press, Cambridge, Mass.

# 2 THE THERAPSIDS

The reign of the pelycosaurs was over before the middle of the Permian, with only a few rather obscure members of the group surviving into late Permian times (Table 1.2). In the Middle Permian they were replaced as the dominant land animals by the members of the Order Therapsida. The latter are classified as an order of reptiles within the sub-class Synapsida, but equate much more closely with a sub-class of the Mammalia, say the Eutheria or placental mammals, than with any individual order of mammals.

## Anomodontia

The therapsids are divided into two sub-orders: the Anomodontia consisting entirely of herbivorous forms and the Theriodontia, which were primarily carnivorous, but produced occasional herbivorous side-branches. The anomodonts are found from the middle of the Permian until the Upper Triassic, a span of $60 \times 10^6$ years (Table 1.2). They were the dominant large to medium-sized land vertebrate herbivores, playing a role very much like that of the ungulate mammals today. Like all such herbivore-assemblages, the anomodonts give rise to no higher group, finally becoming extinct without leaving any descendants. One of the major problems of herbivores is the chopping up and breaking down of vegetable material to a state when their digestive processes can cope with it. Plants are composed of cellulose, which is a tough abrasive material, and they sometimes contain silica as well; both cause considerable wear upon the teeth, and consequently there is a need for continual replacement of the worn out members of the tooth battery. The early anomodonts solved this problem by modifying the cheek-teeth (post-canines) but the later members lost all their teeth except the upper canines and replaced them with a horny beak. A beak is surprisingly efficient for cropping and slicing up plant material, as is demonstrated by present-day herbivorous birds and tortoises. It has the great advantage of continuous growth to compensate for the wear.

The early mid-Permian anomodonts are included in the Dinocephalia (Tapinocephaloidea) and the majority of the described specimens come from the *Tapinocephalus* zone of the South African Karroo, although a number have been described from Russia (Table 1.2). Dinocephalians and the large cotylosaurian parieasaurs were the great herbivores of the Mid-Permian, each competing with the other for dominance; but the latter group soon became extinct and the former were replaced by another group of anomodont reptiles, the Dicynodontia, which formed the dominant herbivores of Upper Permian and Lower Triassic times.

The Upper Permian was the 'hey-day' for the Dicynodontia and it is then that they show the greatest diversity in both size and form. They were also most numerous at this time, however their numbers began to decline in the Lower Trias, as does their diversity and soon no small species are found. It seems that the smaller Upper Permian forms such as *Synostocephalus* failed to compete with the contemporaneous theriodonts and sauropsid reptiles. The large dicynodonts persisted until the Upper Trias, when they too disappeared and with their extinction the anomodonts ceased to inhabit the earth.

## Theriodontia

The other main division of the Therapsida, the Theriodontia contains mammal-like reptiles that were more diverse than the Anomodontia, both in structure and in mode of life. The early theriodonts were carnivorous and this remains their basic habit throughout their history, although some important co-lateral groups did become herbivorous. The earliest theriodonts are the titanosuchids, which were coeval with the earliest anomodonts; the former are sometimes classified in the Dinocephalia along with the tapinocephalids. The resemblances between the two groups are more probably due to parallel evolution and do not indicate any real affinity.

### *Gorgonopsia*

The Gorgonopsia are an important group of Middle and Upper Permian carnivorous theriodonts, which had their 'hey-day' in the latter period. Over seventy genera have been recognised mostly from South Africa and these Romer (1956) classified into twenty families. Sigogneau (1970) much more conservatively classifies the South African gorgonopsids into two families. If this be valid it would make the Gorgonopsia a more diverse group of animals than the mammalian order Carnivora, which has a total of ten families grouped into two sub-orders, the Fissipedia and the Pinnepedia. The Gorgonopsia are admittedly a diverse group but not as much as Romer's classification seems to indicate. They range from animals with skulls a few centimetres long to *Inostrancevia* from the Russian Upper Permian, which had a skull about 480 mm long and was probably one of the largest therapsid carnivores known, being about the size of a grizzly bear. The gorgonopsids were at one time believed to be the ancestors of the later cynodonts (p. 27) but this idea is now less widely held. Despite this the gorgonopsids are worth studying for their own sake, as the fossil material is reasonably plentiful and well-preserved. They have therefore become the best described and best known Permian theriodonts.

As in the Gorgonopsia the definite and characteristic form of the skull of

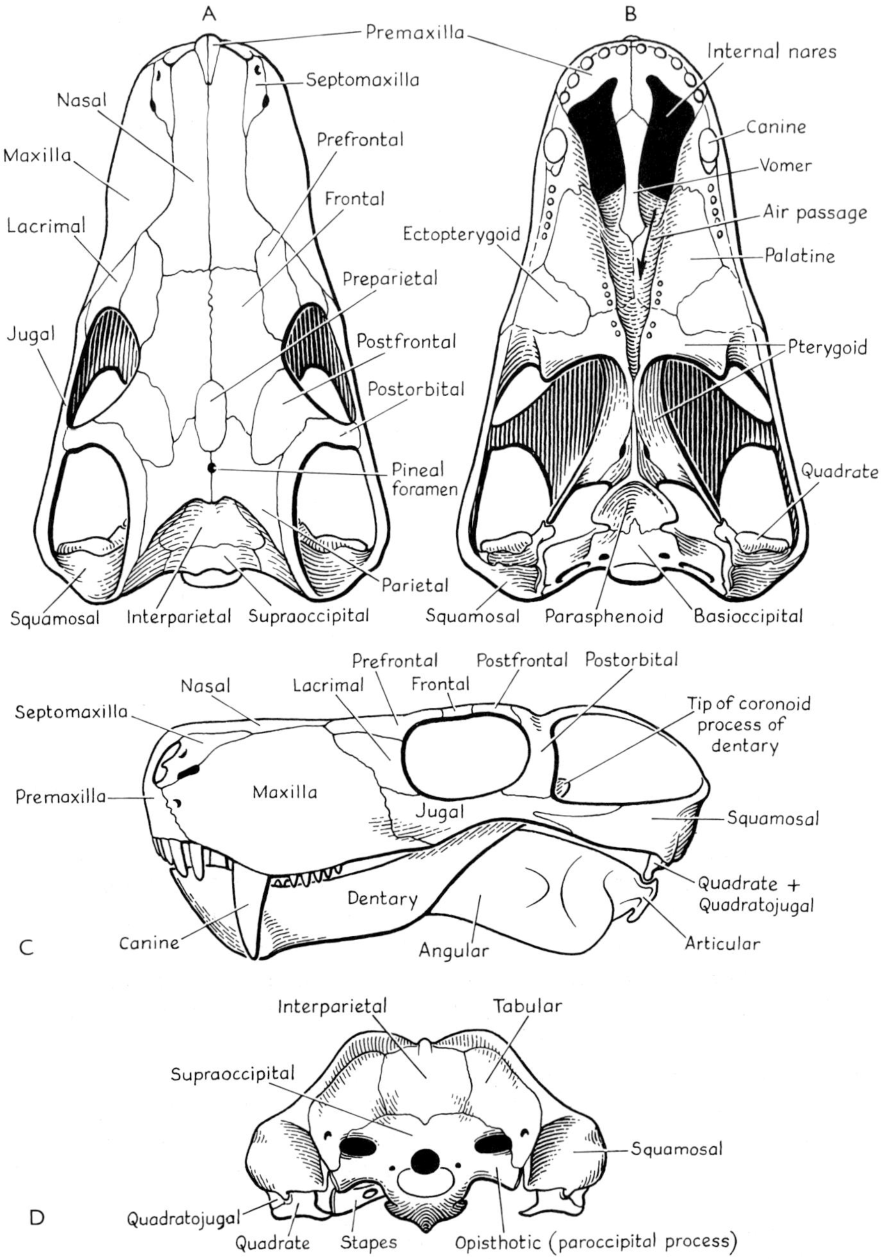


**Figure 2.1:** The Skull of *Dixeya*. A, dorsal view; B, ventral view; C, lateral view and D, occipital view
Source: After Parrington, 1955

the later mammal-like reptiles and eventually the mammals first becomes apparent, the skull of *Dixeya quadrata* will be described in detail (Figure 2.1). This animal is a medium-sized gorgonopsid found in the Upper Permian of Tanzania in beds which are probably equivalent to the *Cistecephalus* zone in South Africa (Parrington, 1955). The skull of *Dixeya* is about 170 mm long. Viewed from above (Figure 2.1A), it shows unusual width between the temporal fossae and the orbits; probably as a direct consequence of this width, a median unpaired bone, the preparietal, lies in front of the pineal foramen. Separating the anterior part of the paired parietals and the posterior part of the paired frontals, this bone is found in typical gorgonopsid skulls. The zygomatic arch is narrow and formed only by the squamosal and jugal, whilst the quadrate is small and fused to the quadratojugal to form one bone, fitting without sutural connection into a small recess in the squamosal. The dental formula of *Dixeya* is $I\frac{5}{4}$, $C\frac{1}{1}$, $PC\frac{5}{5}$, where I = incisors, C = canine and PC = post-canines. The lower jaw of *Dixeya* in lateral view shows that the dentary has a coronoid process as well as a coronoid bone.

In the gorgonopsid palate, the vomers have fused to form a single bone, which ventrally forms a fairly thick horizontal plate expanding posteriorly so it is not visible in Figure 2.1B. This last character is a point of difference between the gorgonopsids and the next group of therapsids to be discussed, the contemporary Therocephalia, where the posterior expansion of the vomer lies on the surface of the palate. In *Dixeya* the single vomer is continued dorsally as a post-card thin lamella, which is continuous with the median ethmoid lamina. The latter is perforated by a large fenestra of unknown function and was first figured by Watson in 1921. These thin laminae were probably covered by a thin sensory epithelium and enabled the animal to smell air-borne scents. This may be the beginning of the development of the acute sense of smell found in many present-day mammals.

The more median parts of the palatines meet the vomer dorsally, so that there is a median trough, rapidly deepening as it passes anteriorly. In life the anterior part of this trough would have been covered by soft tissue to form a functional secondary palate, separating the digestive and respiratory tracts and so enabling the animal to breathe when its mouth was full of food or whilst it was holding on to struggling prey. Although the structure of this secondary palate was different from that found in later mammal-like reptiles and mammals, where it is formed of bone, its function was the same.

The posterior part of the palate is a solid sheet of bone made up of the palatines, ectopterygoids and pterygoids with no fenestrae between them. There is a large medial patch of palatal teeth borne partly by the palatine bones and partly by the pterygoids. In *Dixeya*, the usual row of teeth along each pterygoid flange is absent.

Behind the palate the pterygoids pass backwards and outwards towards the quadrates on either side of the skull. The ventral portion of the vertical rod-like epipterygoid fans out backwards to meet the quadrate and the quadrate ramus of the pterygoid below. The epipterygoid is part of the primary upper jaw structure and therefore lies outside the brain-case with a space, the cavum epitericum between them. Now that its ventral portion is expanded to form a plate, overlying the unossified interorbital region of the brain-case, the epipterygoid is usually called the alisphenoid. Later this bone will become incorporated in the brain-case wall, so that the enlarged mammalian brain is adequately protected. Although the alisphenoid meets the quadrate, the pterygoid stops some distance short of this.

Viewed from the rear, the reduction of the size and height of the quadrate in *Dixeya* has caused a corresponding reduction in the height of the portion of the skull lying below the foramen magnum. Although the occipital condyle is still single and formed mainly from the basioccipital, it is pulled out laterally. The stapes is perforated by an artery and forms contact laterally with the quadrate.

Thus it is with these gorgonopsids that the definite and characteristic form of the skull of the later mammal-like reptiles and eventually the mammals first becomes apparent.

The mounted skeleton of *Lycaenops ornatus* in the American Museum of Natural History Colbert (1948) shows at a glance that the animal had a definite mammal-like stance and appearance (Figure 2.2). The limbs are comparatively long and rotated so that they lie against the body, thus this gorgonopsid had lost completely the sprawling stance and presumably also the lumbering awkward gait of primitive reptiles. The feet lay close to the midline of the animal as they do in present-day mammals, so the animal in life was raised well off the ground and was capable of a precise walk or even of running. Colbert has restored the animal as walking on its toes, so that the foot can be described as digitigrade, as can that of a cat or dog. If he be justified in so doing, the mammal-like appearance of this gorgonopsid is even further enhanced.

With the rotation of the limbs and the consequent change in position of the feet, the outer toes need to become reduced in length for mechanical reasons. There is therefore a reduction in these toes of some of the phalanges to thin discs, and not the actual loss of phalanges as is seen in the anomodonts. In

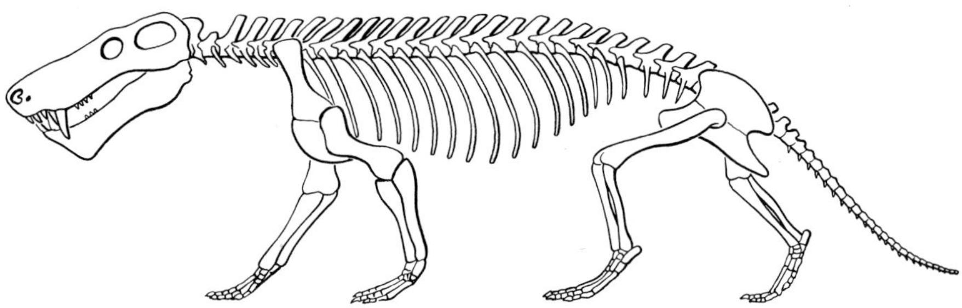

**Figure 2.2:** The Skeleton of *Lycaenops ornatus*
Source: After Colbert, 1948

still retaining the superfluous phalanges, the gorgonopsids are more primitive than the contemporary anomodonts giving an example of the principle that characters do not evolve at the same rate in lineages evolving in parallel.

The tail of *Lycaenops* is much reduced from the pelycosaur condition, which also contributes to this gorgonopsid's mammal-like appearance.

### *Therocephalia*

The Therocephalia are another group of therapsids whose evolution parallels that of the Gorgonopsia. They first appear in the Mid-Permian of Russia and South Africa, with almost all the specimens coming from the *Tapinocephalus* zone of the latter. Their remains are found in deposits in South and East Africa until the Upper Permian and by the end of that period most of them have disappeared.

The therocephalians were not such large reptiles as the gorgonopsids, neither do they show the extreme development of the upper canines. *Lycosuchus* along with other early therocephalians has a pair of functional canines in the maxilla. This is a 'carry-over' of the primitive condition found in the Pelycosauria but which had been lost in the titanosuchids and gorgonopsids, both of them contemporaries of the early therocephalians. Despite the presence of this primitive character, there is a powerful coronoid process to the dentary of the lower jaw for the attachment of muscles. The palate of *Scylacosaurus* shows that in this genus there was just one functional canine on each side and in general appearance it resembles that of a gorgonopsid (Figure 2.3). There is, however, one marked difference and that concerns the position of the posterior extension of the vomer. In gorgonopsids it is above (i.e. dorsal) to the palatines, whilst it is below them in not only the therocephalians but also in their possible descendants, the cynodonts (Boonstra, 1934a, b). In addition the palate of these early therocephalians can show that at least the anterior part of the vomer is formed of two unfused bones. This is a primitive feature that the Gorgonopsia have lost; similarly the early therocephalians do not have the incipient secondary palate. Instead of the solid sheet-like posterior part of the palate, there is a large vacuity on either side lying between the palatine, ectopterygoid and the pterygoid in *Scylacosaurus*.

The posterior part of the skull is similar to that of a gorgonopsid, with one exception; namely the ascending ramus of the epipterygoid (alisphenoid) has expanded in the Therocephalia to form a narrow plate, as against a narrow vertical strut. In other words, the alisphenoid in therocephalians is beginning to look like that found in the mammalian skull, but it lies lateral to and has no connection with the lateral wall of the brain-case proper. The space between the two is the cavum epiptericum and in it lies the ganglion of the fifth cranial nerve.

The post-cranial skeleton of the therocephalians is not well-known, but

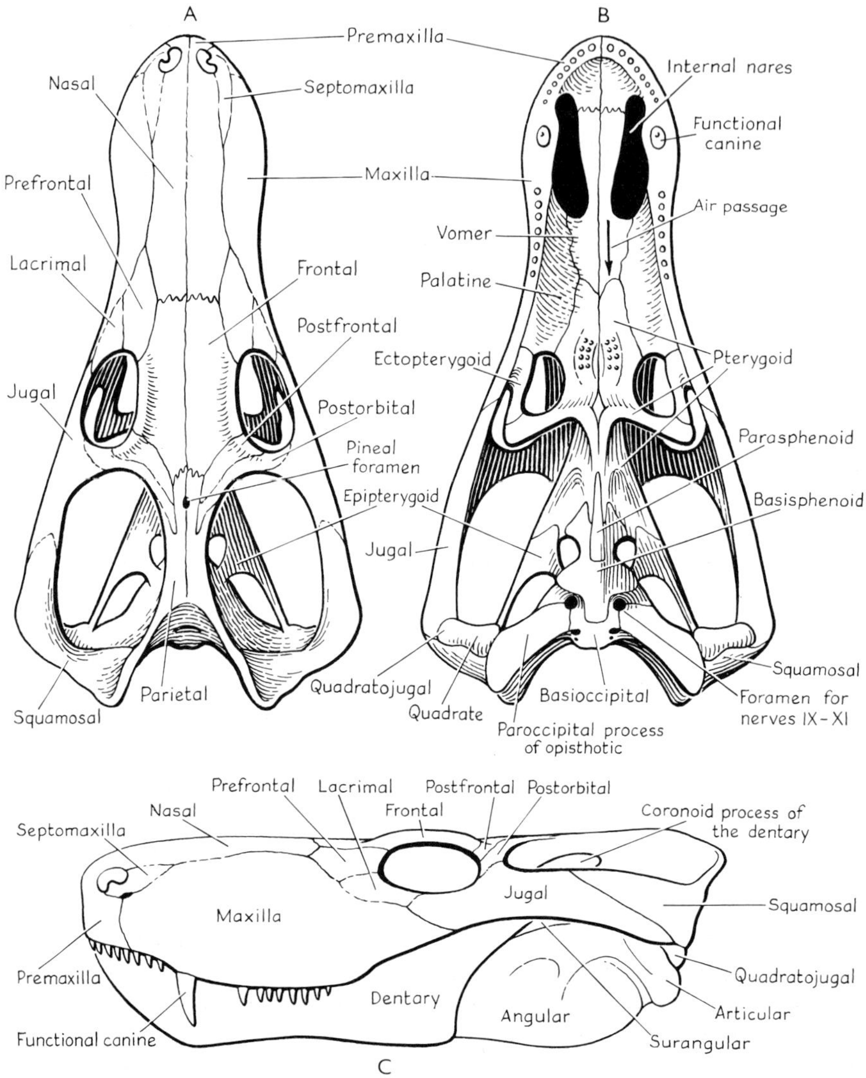


**Figure 2.3:** The Skull of *Scylacosaurus*. A, dorsal view; B, ventral view and C, lateral view
Source: After Boonstra, 1934b

indications are that it resembled that of the gorgonopsids. However, in some Therocephalia, the phalangeal formula is reduced to the mammalian 2.3.3.3.3.

*Whaitsia* and its relatives, found at the end of the Upper Permian (mostly *Cistecephalus* zone of South Africa) are a later and more specialised group of Therocephalia, forming a prematurely specialised and sterile side-branch from the main stock. None survive the Permian and they die out leaving no descendants. In these animals the skull is low and broad, and all the post-canine teeth have been lost. The palate is deeply troughed and a functional secondary palate analogous to that found in the gorgonopsids has been developed; this is yet another example of parallel evolution in the different groups of the Therapsida. In some whaitsiids, the maxillae may actually join along the mid-line to give a true bony palate. The palatal fenestrae found in some of the more primitive Therocephalia are reduced or absent in the whaitsiids.

### *Bauriomorpha*

Romer (1956) places the more advanced Therocephalia in a separate infraorder, the Bauriomorpha. This has the advantage of dividing a large and heterogenous group into earlier more primitive forms (the Therocephalia proper) and the later, more advanced forms (the Bauriomorpha). Unfortunately it has the disadvantage of making a sharp cleavage where in fact none may exist; also it obscures relationships by arbitrarily cutting across phyletic lines. More needs to be known about the therapsids in the wide sense, their relationships and their evolution before a reasoned discussion can be made about even the broad basis of their classification.

Of the Bauriomorpha, using Romer's term without prejudice, one genus, *Bauria* itself, will be briefly described. It comes from the top of the South African Trias in the *Cynognathus* zone and with a skull some 130 mm long was one of the largest bauriomorphs. The dorsal view of the skull shows that the pineal foramen was closed and the post-orbital bar incomplete (Figure 2.4A and 2.4C). The ascending process of the epipterygoid is a narrow vertical strip of bone and therefore shows no advance over the condition found in Permian therapsids. The palate (Figure 2.4B) is rather primitive, with the palatines apparently taking no part in the formation of a secondary palate. Posteriorly there are large fenestrae lying between the pterygoids, ectopterygoids and palatines; these are so typical of all Therocephalia, using that term in its widest sense. There is a single unpaired occipital condyle formed mainly from the basioccipital (Figure 2.4D). The lateral view of the skull shows the simple peg-like cheek-teeth, so characteristic of the Bauriomorpha (Figure 2.4C). The complex post-canines with accessory cusps found in many of the other great groups of Lower Triassic therapsids, e.g. the cynodonts, are completely lacking. In addition the structure of the lower jaw with its large post-dentary bones resembles the condition found in a gorgonopsid or an early therocephalian.

*Bauria* is remarkable in showing the juxtaposition of primitive and advanced characters. The closed pineal foramen and incomplete post-orbital bar contrast strangely with the simple peg-like teeth, rather primitive secondary palate, well-developed post-dentary bones and single occipital condyle.

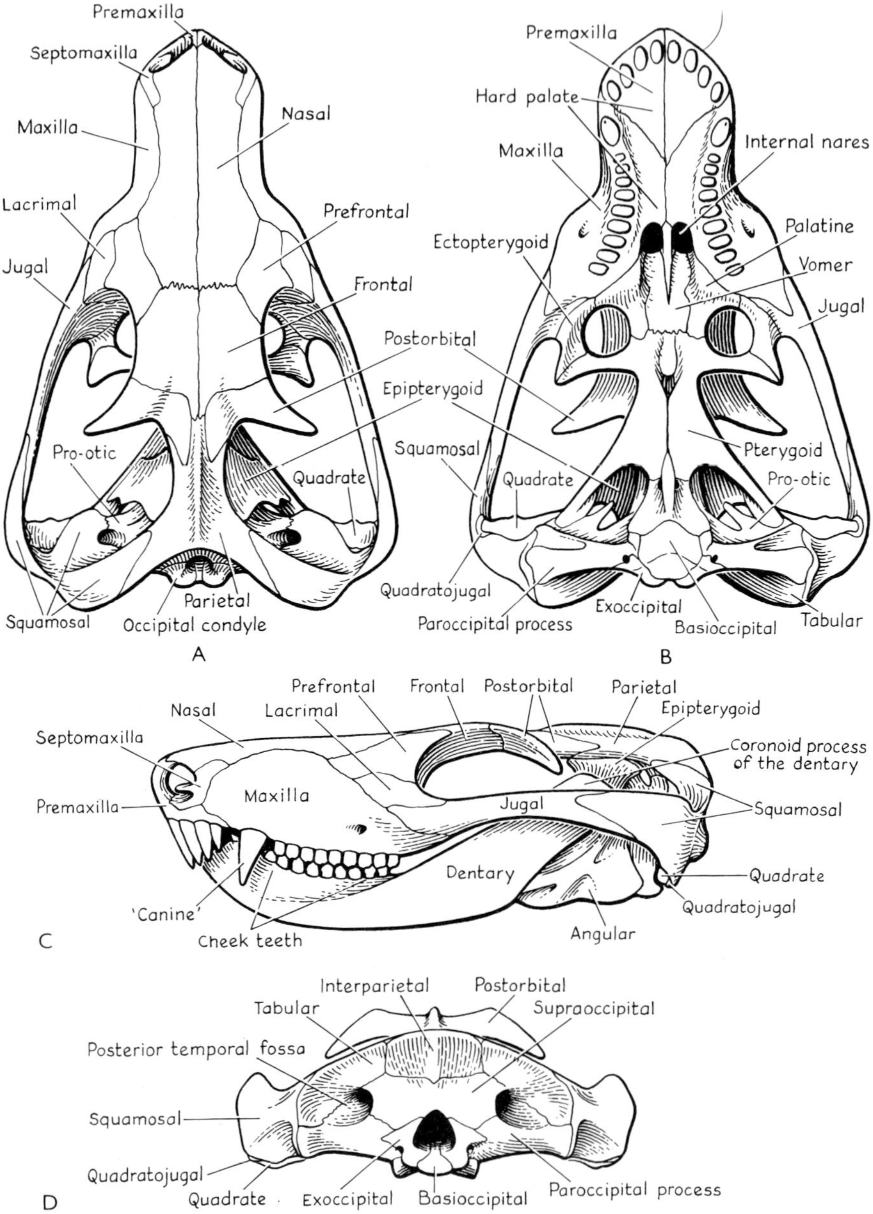


**Figure 2.4:** The Skull of *Bauria*. A, dorsal view; B, ventral view; C, lateral view and D, occipital view
Source: After Brink, 1963a

*Cynodontia*

The Cynodontia are the most important and largest group of Triassic Therapsida. The earliest ones come from the Upper Permian of South Africa (*Cistecephalus* zone) and from Russia. The last cynodont, *Scalenodontoides*, is found in deposits in the lower part of the Upper Trias of Lesotho (Upper Molteno). Which of the more primitive groups of Therapsida was ancestral to the cynodonts is still disputed. Watson (1921) held that the Gorgonopsia were the basal group and from them the Cynodontia and the Therocephalia evolved independently. His main grounds for deriving the cynodonts from the gorgonopsids were the absence of palatal vacuities and the persistence of the phalangeal formula 2.3.4.5.3. in both groups. It seemed that in some groups of early Therocephalia the reduction to the mammalian formula of 2.3.3.3.3. had already taken place. Initially Broom (1932) was prepared to accept Watson's theory.

In 1934 Boonstra showed that the posterior expansion of the vomer lies dorsal to the palatines in Gorgonopsia and ventral to them in the Therocephalia. Cynodonts resemble the therocephalians in this respect rather than the gorgonopsids, and the condition of the palate in the former cannot be derived from that in the latter and *vice versa*. This fundamental difference in the structure of the palate is evidence for the Cynodontia having evolved from the Therocephalia and not from the Gorgonopsia.

In 1938, Broom published a paper on the origin of the cynodonts, in which he described two new primitive forms from the Upper Permian (*Cistecephalus* zone) of the Karroo of South Africa. Broom considers the general therocephalian appearance of these two skulls, both species of *Procynosuchus*, sufficient to favour an origin of the Cynodontia from the Therocephalia rather than from the Gorgonopsia. Rather surprisingly Broom does not mention the critical observation by Boonstra on the palatal structure which is the strongest evidence of the therocephalian origin for the cynodonts. The question is still open, but at present the majority opinion favours the origin of the Cynodontia from the Therocephalia (Kemp, 1982). Romer (1969) however, considers that the cynodonts arose independently of either Gorgonopsia or Therocephalia. We incline to this view.

The cynodonts are a large and well-known group and many skulls have been described and figured, but unfortunately most of the descriptions confine themselves to the superficial bones, with the result that the braincases have not been well known. It is for this reason, that a representative of a late group of cynodonts will be described, as in *Diademodon (Gomphognathus)* the braincase is known in some detail. The Diademodontidae are the last group of cynodonts to become extinct and they can be regarded as the most specialised of all. Whilst most cynodonts, and certainly all the early ones, were carnivorous the Diademodontidae had become modified for a herbivorous or at least, an omnivorous diet. In this respect they resemble present-day bears and badgers amongst the Carnivora. Plant-eating always

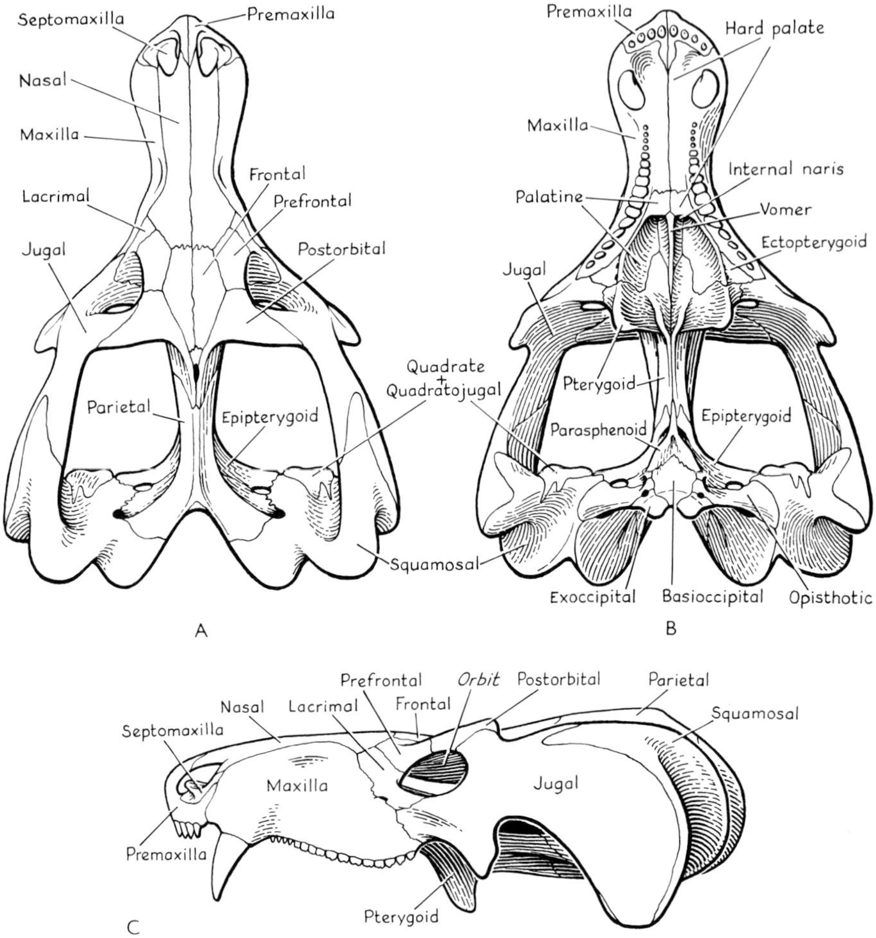


**Figure 2.5:** The Skull of *Diademodon*. A, dorsal view; B, ventral view and C, lateral view
Source: After Brink, 1963b

produces considerable modification of the dentition and jaws and so *Diademodon* and its allies are rather atypical of cynodonts in this respect. This should be borne in mind throughout the following account.

Figure 2.5 illustrates the great width of the skull across the zygomatic arches and their powerful development. As the braincase itself is narrow, most of the width across the skull is due to the enormously expanded lateral temporal fossae. Each of these becomes so large as to break through to the parietal dorsally, so that bone and not the squamosal and postorbital forms its upper boundary. In *Cynognathus*, a carnivorous cynodont contemporary with *Diademodon*, the relationship of the lateral temporal fossae to the skull bones has become even more aberrant. Not only has each fossa 'broken through' dorsally to the parietal, but the postorbital has grown back to meet the squamosal and so exclude the jugal from the lower border of the fossa. This is associated with the increase in depth of the zygomatic arch shown by many later cynodonts.

In *Diademodon*, the space between the braincase and the zygomatic arch housed the muscles which closed the lower jaw. These muscles must have been extraordinarily powerful, as is also indicated by the great size of their areas of attachment on the deep zygomatic arches and enlarged sagittal and occipital crests of the braincase.

The lateral view of the skull (Figure 2.5C) shows the general flattening as well as the enormous development of the zygomatic arch. The area available for the attachment of the jaw muscles is further increased by the downwardly projecting process from the arch just below the orbit. The well-developed pterygoid flanges to locate the lower jaw and prevent its displacement sideways by the action of the powerful jaw muscles can also be seen in this view.

There is a toothless gap or diastema between the canines and the cheek-teeth. Such a diastema is found today in herbivorous mammals and is one of the grounds for considering *Diademodon* and its relatives as herbivores.

The ventral view of the skull (Figure 2.5B) shows the true bony secondary palate formed mainly by the ingrowth of the maxillae to form a flange and also by the growth inwards of the palatines posteriorly. This secondary palate does not extend quite so far back as in true mammals, but in all other respects it is identical. This view of the skull also shows the powerful pterygoid flanges and the absence of any palatal teeth such as are found in the skulls of earlier therapsids. Like true mammals, *Diademodon* and its allies had their teeth confined to the margins of the jaw and borne on the premaxilla, maxilla and dentary. Thus the skull of *Diademodon* shows a considerable advance towards the mammalian condition from that found in the earlier therapsids.

In *Diademodon* the quadrate forms the skull portion of the hinge with the lower jaw and it can be seen in Figure 2.5A and B. It is a small bone fused laterally with the quadrato-jugal; the combined bone fits into a recess in the squamosal but forms no sutural connection with it. Being loose, this quadrate-complex is easily lost and so it will only be found in skulls which are fossilised with the lower jaw in position. In reptiles the quadrate forms part of the jaw joint, whereas in typical mammals it has become free and forms one of the bones, the incus, of the middle ear, which transmits sound waves from the ear-drum to the inner ear. The ear-drum of *Diademodon* was attached to two bones of the lower jaw, the angular and articular. It was not attached to the quadrate. The objection has often been made that, if the tympanic membrane or ear-drum were attached in this way to the bones of the lower jaw, the animal would deafen itself every time it opened its mouth. This objection is quite fallacious. In present day lizards, the tympanic membrane is attached along its anterior border to the quadrate, which in most lizards is moveable. They appear to manage quite well. Certainly in an analogous condition using electronic equipment, there would be no difficulty in designing a 'blocking circuit' to prevent a loud bang being heard every time the animal opens its mouth.

The stapes butted against the quadrate and was a strut of bone perforated by the stapedial artery and orientated with its greatest width running vertically. Thus when viewed from above, particularly in an inadequately prepared specimen, it might give the impression of being a bony rod and its true nature obscured. The proximal end of the stapes would terminate in the foot-plate, which fitted into the fenestra ovalis and rested on the membrane which closed this orifice. Thus, by means of the articular, quadrate and stapes sound waves could be transmitted from the tympanic membrane in the lower jaw to the inner ear, where they were would be transformed into nervous impulses (see Chapter 6).

Returning to Figure 2.5B showing the ventral view of the skull, this shows the extreme narrowness of the anterior part of the braincase. So narrow is this part of the braincase, that it is quite inconceivable that any part of the brain extended forward into it; there simply would not have been room. The parasphenoid bone forms a strongly developed ridge in the lower portion of this anterior region and here the pterygoid forms the most ventral part of the lateral wall with the major part being formed by the alisphenoid. More posteriorly, the pterygoid and alisphenoid diverge from the wall of the braincase and run laterally to meet the quadrate and squamosal. In this posterior region, the braincase has become considerably wider and housed the brain, with its floor being formed by the basisphenoid and the basioccipital, and its lateral walls on each side by the bones which house the inner ear, namely the pro-otic and opisthotic. These latter bones were formed by the ossification in the originally cartilaginous neurocranium, whilst the alisphenoid is by origin a palatal bone. Thus the alisphenoid does not form part of the braincase wall, except in its most anterior part. In the more primitive therapsids, the alisphenoid, here more usually called by its reptilian name of epipterygoid, is a narrow vertical strut of bone lying laterally clear of the braincase wall. In the eutherian mammals (marsupials and placentals) the alisphenoid forms the lateral wall of the braincase and it is often said that this condition had already been acquired in forms at the level of evolution of *Diademodon*. Figure 2.6A showing the lateral view of the braincase with the zygomatic arch removed would seem to support this hypothesis, since a large part of the lateral wall of the braincase appears to be composed of the alisphenoid. However Figure 2.6B, which is a vertical sagittal half of the skull showing the inside of the braincase wall, gives quite a different picture. Here the posterior part of the braincase is formed by the pro-otic and the opisthotic. The discrepancy is explained by recalling that the posterior part of the alisphenoid and pterygoid passed lateral to the braincase wall and was separated from it by a space, the cavum epiptericum, which lodged amongst other structures the semilunar or gasserian ganglion of the fifth or trigeminal

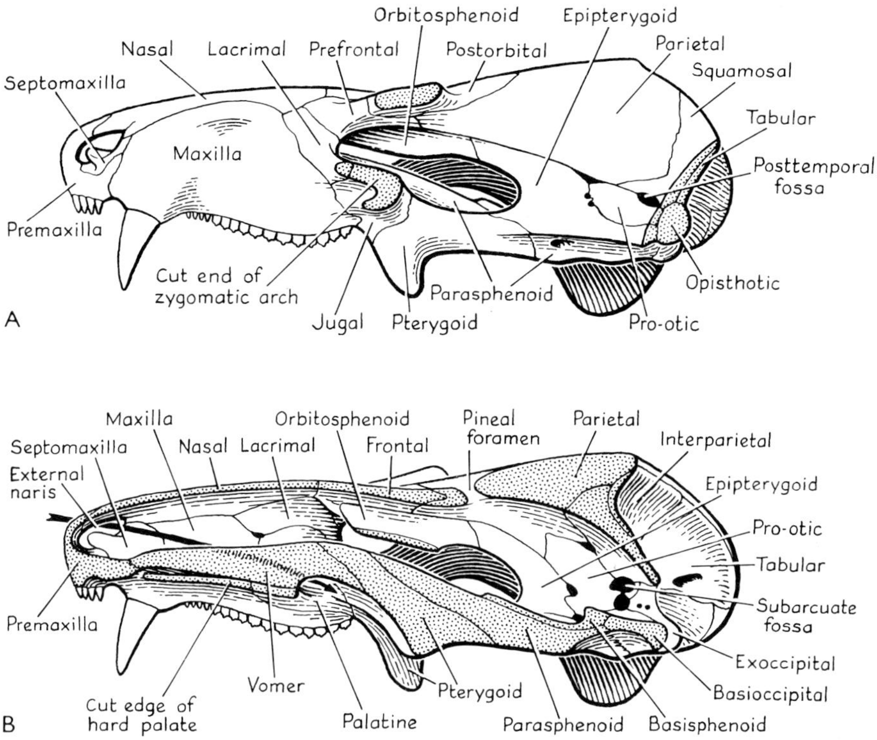


**Figure 2.6:** Lateral Views of the Skull of *Diademodon*. A, with the zygomatic arch removed; B, with the left side of the skull removed
Source: After Brink, 1955

nerve. In the Eutheria this ganglion lies within the braincase, which has clearly incorporated the cavum epiptericum, thus giving the extra space needed to house the greatly enlarged brain.

The sagittal section of the skull is worth a little more study. Between the opisthotic and the pro-otic is a large space which housed the major part of the inner ear and posterior to the latter are two foramina, which were for the passage of the XIIth or last cranial nerve. The facial or VIIth cranial nerve passed through the pro-otic in front of the inner ear and was the nerve which innervated the face. In modern mammals with their fleshy lips, mobile snout and ears, the facial muscles are large and important, so the facial nerve is correspondingly large. In a modern reptile there is little facial muscle and likewise the VIIth nerve is small. The size of this nerve in *Diademodon* corresponds more closely to that of a modern crocodile than it does to that of any modern mammal. Thus *Diademodon* probably did not have the expressive face of a dog or pig, but more likely the 'blank expression' of a crocodile or lizard.

Both Watson (1913) and Brink (1955) state that the ventral (internal) surfaces of the nasals are grooved and they suggest that in *Diademodon* this grooving indicates the presence of turbinal bones like those in the nasal passage of a mammal. In the latter, the function of the turbinals is to support the many folds of the sensory mucous membrane, over which inspired air is drawn, warmed and freed from dust before being taken into the lungs. The membrane also has an olfactory or smelling function as minute olfactory sense organs are incorporated in it. Brink suggests that the probable presence of the turbinal bones in *Diademodon* is an indication that these forms were 'warm-blooded'. Following the same line of thought he comments (1955) on the presence of a deep depression with anteriorly running grooves in the external surface of the maxilla. This he considers lodged a gland and he suggests that it supports the idea of a fleshy snout in this cynodont. This may be so, but the small size of the facial nerve in *Diademodon*, all late therapsids and early mammals as well, suggests the exact opposite. The modern pit-vipers (Crotalinae) have a similar deep pit in their maxillae, which lodges an infra-red detector but they do not have a fleshy snout and the overall appearance of their skulls is anything but mammalian. The function of the pit in *Diademodon* remains a mystery.

The lower jaw of *Diademodon* is massive and when viewed laterally, virtually the whole surface is formed of one tooth-bearing bone, the dentary (Figure 2.7A). The only other bone to be seen in this aspect is the articular located at the posterior end of the dentary and forming the hinge with the quadrate of the skull. Running forward from the hinge region is a well-developed ridge on the surface of the dentary, which has two functions, one to strengthen the jaw and the other to give attachment to the masseter muscle. A great coronoid process on the dentary gives an additional area for the insertions of the masseter and temporalis muscles used in opening and closing the jaw. Below the ridge, the posterior edge of the dentary falls sharply to form a well-marked obtuse angle with its ventral edge.

In lateral view the lower jaw has a definite mammalian appearance, not only in its general shape, but also in the presence of a diastema and the broadened crowns of the cheek-teeth. Only the small articular at the hinge is a reminder that the animal is really a reptile.

Figure 2.7C shows a medial view of the dentary alone, the other bones having been removed. At the posterior end below the coronoid process is a deep trough, which approximately corresponds in position to the ridge seen on the lateral side of the jaw. A powerful ridge or shelf runs dorsally to the trough. This ends in a foramen, the mandibular foramen, through which the inferior dental nerve (a branch of the Vth or trigeminal nerve) enters the dentary. From this point a groove (the internal mandibular groove) runs forward along the dentary to the symphysis. It is covered by the splenial bone, and housed a nerve and an artery.

Figure 2.7B shows the complete lower jaw with the reptilian jaw bones lying in the trough with the shelf above giving them additional support. The

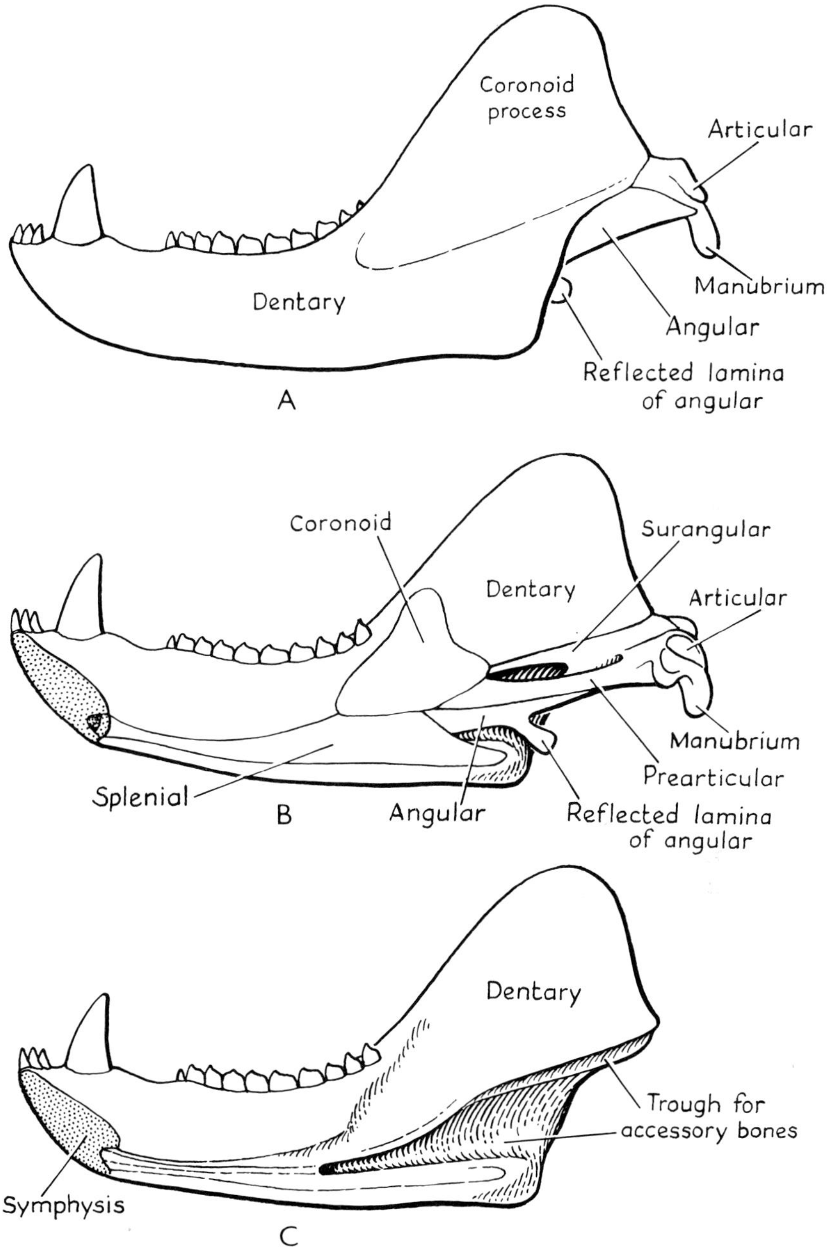


**Figure 2.7:** The Lower Jaw of *Diademodon*. A, lateral view; B, medial view and C, medial view with the accessory jaw bones removed
Source: After Crompton, 1963

most posterior bone of the jaw is the articular, which with the quadrate on the skull, forms the hinge of the jaw. The articular has a well-developed retro-articular process, which will be discussed later in connection with the ear (Chapter 6). Other bones lying in the trough are: the prearticular, which, although it seems to be a forward extension of the articular is not, like the latter, an ossification of Meckel's cartilage; the surangular, which lies dorsal to the articular-prearticular; and the angular which lies ventral to them. The angular bears a backwardly directed process, the reflected lamina. This will be referred to again in the chapter on hearing (Chapter 6). Finally there are two bones lying outside the trough altogether; these are the splenial and the coronoid. This last bone lying dorsally at the posterior end of the groove functions as a bearing surface for the pterygoid flange and to prevent lateral displacement of the lower jaw.

The general appearance of the post-cranial skeleton of *Diademodon* indicates that it was a powerfully built but rather squat animal, with a long body, short legs and tail. The legs may have splayed out to some extent. The anterior ribs of the body were long and powerful, but in the posterior part of the body they suddenly shorten, so that there is a short lumbar region without a rib-cage (Figure 2.8).

In mammals there is a similar division of the trunk into two parts: an anterior part, the thorax, in which the ribs form a flexible bony cage; and a posterior part, the abdomen, in which the ribs are short. The division between thorax and abdomen is marked by the diaphragm, which is a great sheet of muscle running transversely and dividing the body cavity into an anterior thoracic cavity and a posterior abdominal one. The former contains the lungs and the heart; the latter the liver, kidneys and the major part of the gut. The inflation and deflation of the lungs in breathing is brought about by the contraction and relaxation of the diaphragm, aided by the intercostal muscles and the rib cage. It is not known if *Diademodon* and the later cynodonts had a diaphragm and so breathed in the same way as the mammals.

According to Brink (1955), the lumbar ribs in *Diademodon* overlap and

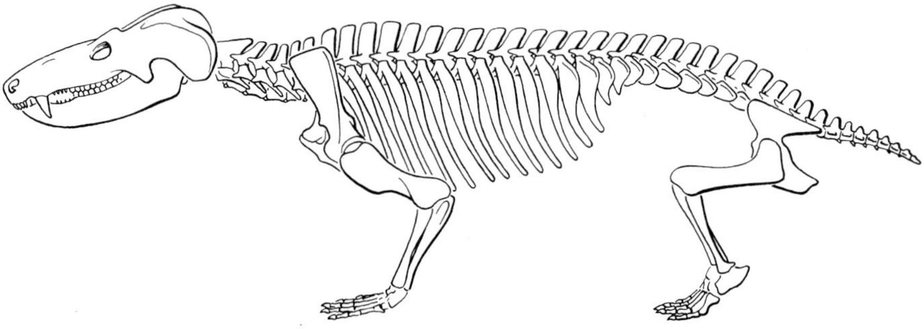

**Figure 2.8:** The Skeleton of *Diademodon*
Source: After Brink, 1955

clasp one another so as to prevent any rotation of the vertebral column. Jenkins (1971) has shown an analogous state of affairs in *Thrinaxodon*, and the condition seems to be common to most, if not all cynodonts. This contrasts strongly with the condition found in modern carnivores, in which the lumbar region allows great rotation. This can be demonstrated with the co-operation of a friendly cat. No matter how much the cynodonts, and in particular the smaller carnivorous ones such as *Thrinaxodon*, appear superficially mammalian in appearance; they would seem very stiff and clumsy, quite lacking the flexibility and grace of a cat or a dog, if we could see them alive today.

In *Diademodon* (Brink, 1955) the phalangeal formula of the hand and foot is the mammalian 2.3.3.3.3, whereas in other cynodonts such as *Thrinaxodon* (Parrington, 1939) the number of phalanges is intermediate between the reptilian formula of 2.3.4.5.4 and the mammalian one. However, in the manus of *Thrinaxodon*, while the phalangeal formula is 2.3.4.4.3, the second phalanx in both the third and fourth digits has been reduced to a mere disc. Thus the *functional* formula in *Thrinaxodon* was the mammalian one, with the supernumerary phalanges undergoing drastic reduction.

Mammals are distinguished from reptiles by maintaining a reasonably constant body temperature, which is almost always higher than that of the environment. This is true even of the most primitive living mammals, the spiny ant-eaters (*Tachyglossus* and *Zaglossus*) and the duck-bill platypus (*Ornithorhynchus*) of the Australian Region. In these animals the body temperature is by no means as constant as it is in placental mammals, fluctuating over fairly wide limits; but it is better regulated than that of any living reptile. Sometime the 'cold-blooded' (poikilothermic) ancestors of the mammals must have start to evolve towards homiothermy. Whether these first steps towards 'warm-bloodedness' were first taken by the later cynodonts is not and cannot be known. A regulated, raised body temperature may not have appeared in the mammalian stock until much later.

The cynodonts, lasting as they did some $45 \times 10^6$ years from the Upper Permian to the Upper Trias, provide an interesting chapter in the history of the ancestors of the mammals (Table 1.2). As a group they made great advances in perfecting their dentition, jaw articulation and hearing, while retaining many primitive reptilian characters. But they failed in the end to withstand the competition of the newly evolving groups of the sauropsid reptiles, in particular the dinosaurs, and consequently became extinct. The future was with another group of the Synapsida — the mammals.

## References

Boonstra, L. D. (1934a) 'Additions to our Knowledge of the South African Gorgonopsia, preserved in the British Museum (Natural History)', *Ann. S. Afri. Museum, 31*, 177–213

Boonstra, L. D. (1934b) 'A Contribution to the Morphology of the Mammal-like Reptiles of the Suborder Therocephalia', *Ann. S. Afri. Museum, 31*, 215–267

Brink, A. S. (1955) 'A study on the Skeleton of *Diademodon', Palaeontologica Africana, 3*, 3–39

Brink, A. S. (1963a) 'On *Bauria cynops* Broom', *Palaeontologica Africana, 8*, 39–56

Brink, A. S. (1963b) 'Notes on Some New *Diademodon* Specimens in the Collection of the Bernard Price Institute', *Palaeontologica Africana, 8*, 97–118

Broom, R. (1932) *The Mammal-like Reptiles of South Africa and the Origin of Mammals,* H. F. and G. Witherby, London

Broom, R. (1938) 'The Origin of the Cynodonts', *Ann. Transv. Mus., 19*, 279–288

Colbert, E. H. (1948) 'The Mammal-like Reptile *Lycaenops', Bull. Amer. Mus. Nat. Hist., 89*, 355–404

Crompton, A. W. (1963) 'On the Lower Jaw of *Diarthrognathus* and the Origin of the Mammalian Lower Jaw', *Proc. Zool. Soc. Lond., 140*, 697–753

Jenkins, F. A. (1971) 'The Postcranial Skeleton of African Cynodonts', *Bull. Peabody Mus. Nat. Hist., 36*, 1–216

Kemp, T. S. (1982) *Mammal-like Reptiles and the Origin of Mammals,* Academic Press, London

Parrington, F. R. (1939) 'On the Digital Formulae of Theriodont Reptiles', *Ann. Mag. Nat. Hist.,* ser.11, *3*, 209–214

Parrington, F. R. (1955) 'On the Cranial Anatomy of Some Gorgonopsids and the Synapsid Middle Ear', *Proc. Zool. Soc. Lond., 125*, 1–40

Romer, A. S. (1956) *Osteology of the Reptiles*, The University of Chicago Press, Chicago

Romer, A. S. (1969) 'The Chañares (Argentina) Triassic Reptile Fauna V. A New Chiniquodontid Cynodont, *Probelesodon lewisi* — Cynodont Ancestry', *Brevoria, 333*, 1–24

Sigogneau, D. (1970) '*Révision Systématique des Gorgonopsiens Sud-Africains*', Editions du Centre National de la Recherche Scientifique, Paris

Watson, D. M. S. (1913) 'Further Notes on the Skull, Brain, and Organs of Special Sense of *Diademodon*', *Ann. Mag. Nat. Hist.,* ser. 8, *12*, 217–228

Watson, D. M. S. (1921) 'The Bases of Classification of the Theriodontia', *Proc. Zool. Soc. Lond.*, pp. 35–98

# 3 SPECIALISED CYNODONT DERIVATIVES

Under this heading several lineages arising from the advanced cynodonts will be considered. These lines do not appear to have any close relationship to each other, but they all in some way transcend the typical cynodont level of evolution. All the lines eventually became extinct, and only one (the Haramiyidae), in our opinion, can be regarded as possible ancestors of any of the groups of mammals, although in some features they parallel the early mammals closely. Once again, with similar problems to be solved similar solutions were produced. The four lineages are:

(1) Chiniquodontidae,
(2) Ictidosauria,
(3) Tritylodontidae,
(4) Haramiyidae.

## Chiniquodontidae

The Chiniquodontidae was an advanced family of carnivorous cynodonts from the Middle Triassic of South America (Chañares and Santa Maria formations). There are four genera: *Belesodon, Chiniquodon, Probainognathus* and *Probelesodon*. The first two were described by von Huene (1942) from the Santa Maria formation of Brazil. *Probainognathus* and *Probelesodon* have both been described by Romer (1969, 1970, 1973) from the Chañares Formation in Argentina. These Chañares beds are slightly earlier than the Santa Maria. Although good material is known of both these genera, it is the dentition and the jaw articulation which is of particular interest here.

The jaw articulation in *Probelesodon* seems to have evolved from an articulation similar to that described by Crompton (1972) in the cynodont *Thrinaxodon*. The surangular has acquired a contact with the squamosal. This formed a thrust-bearing which resisted the couple tending to dislocate the jaw when the cheek-teeth came into occlusion. As is typical in cynodonts, the cheek-teeth in *Probelesodon* do not seem to have had the typical definite mammalian occlusal pattern, nor do they show heavy wear.

*Probainognathus* was a much smaller animal and the jaw articulation is dramatically different (Figure 3.1). The end of the surangular forms a small condyle which articulates with a small, but quite definite glenoid on the squamosal. There may have been a slight contact between the dentary and the squamosal. Here there is a much more powerful thrust-bearing than in any

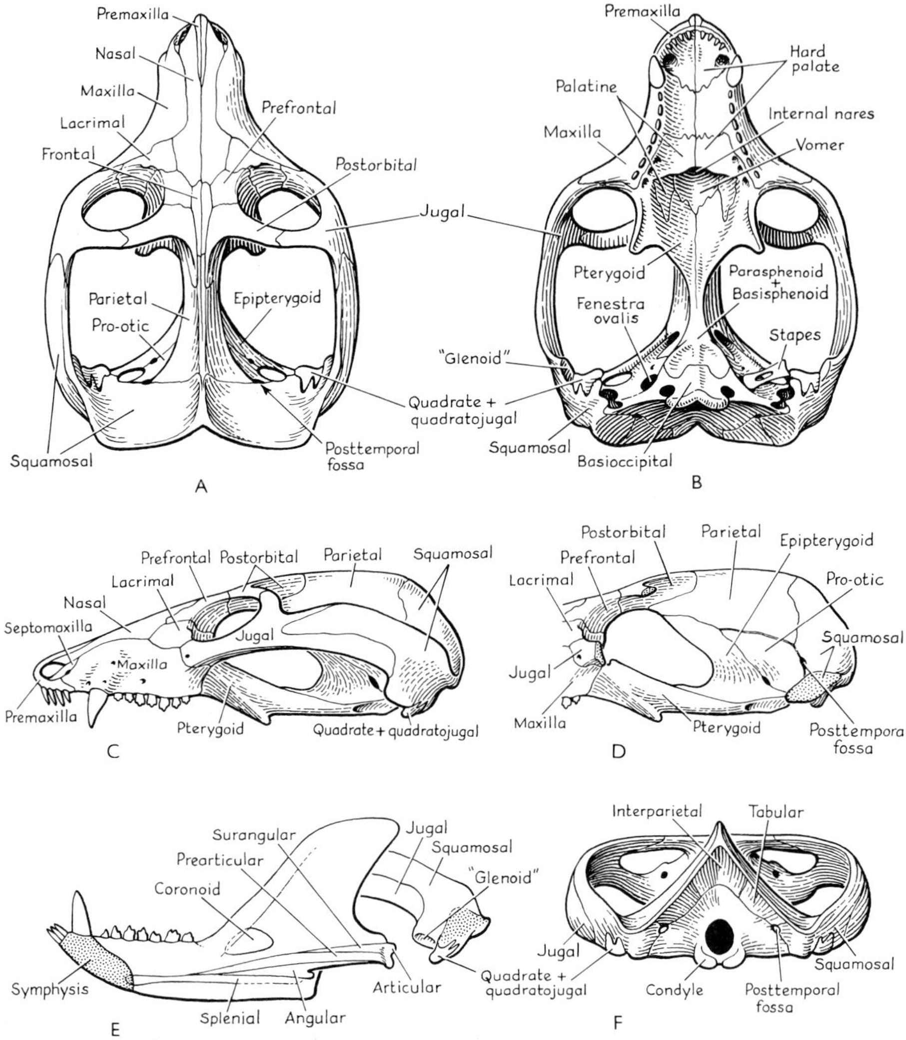


**Figure 3.1:** The Skull of *Probainognathus*. A, dorsal view; B, ventral view; C, lateral view and D, lateral view of braincase with zygomatic arch removed; E, medial view of lower jaw and posterior end of zygomatic arch and F, occipital view
Source: After Romer, 1970

other cynodont. Uniquely amongst cynodonts, the post-canine teeth show heavy wear and came into precise occlusal relationships as do the equivalent teeth in a mammal. During use, the cheek-teeth of *Probainognathus* must have been held in active occlusion by the jaw muscles in a typically mammalian way; in addition, they must have undergone active sharpening by tooth to tooth contact (thegosis), which is equally typical of the Mammalia.

The surangular-squamosal articulation and the possible dentary-squamosal contact persuaded Romer that *Probainognathus* lay on the direct line of evolution of the mammals, and in this he was followed by Crompton (1972). However, we are certain that the resemblances of jaw-joint and teeth in *Probainognathus* and mammals is the result of parallel evolution. Similar selection pressures have led to the evolution of similar morphological structures. In almost all cranial characters, and in particular the structure of the braincase, *Probainognathus* is a typical advanced cynodont. Advanced cynodonts have specialisations (apomorphic characters) which do not occur in *Morganucodon*, the earliest mammal whose skull is known (Kermack, Mussett and Rigney, 1981 — see Chapter 4). The most notable of these apomorphic characters in the cynodonts is the greatly broadened epipterygoid, which contrasts with the primitive narrow, almost gorgonopsid-like, epipterygoid of *Morganucodon*. Thus the chiniquodonts and the morganucodonts stand next to each other in the relationship of sister groups; one is not the ancestor of the other.

## Ictidosauria

Ictidosauria was the term first applied by Broom (1932) to two small reptiles discovered in the 1920s by van Hoepen in the Cave Sandstone of Ladybrand in the Orange Free State, South Africa. The Cave Sandstone is of Rhaetian or Lower Jurassic age (Table 1.2). Van Hoepen intended to name and describe these specimens himself, but he permitted Broom to examine them and publish a brief description. Broom was mostly interested at that time in the origin of the mammalian vomer, but he commented on the generally mammalian appearance of the two skulls, which he referred to as Ictidosaur A and B. In particular he said 'The dentary could easily form a new joint with the squamosal, which it already must touch'. Van Hoepen never did describe or name his specimens, and there the matter rested until he retired from the Directorship of the National Museum in Bloemfontein after the Second World War. Then Crompton (1958) redescribed Ictidosaur B and stated that there was indeed a squamosal-dentary contact. The dentition, however, in no way resembled that of the mammals. He renamed the specimen *Diarthrognathus broomi*, but it was generally agreed that it was not a mammal or a mammal ancestor, but a sterile side-branch of the cynodonts, like *Probainognathus*. Recently Gow (1980) has doubted the existence of the squamosal dentary contact. From a brief examination one of us gave the specimen many years ago we would agree with Gow.

Often included in the Ictidosauria are two small forms from the southern African Red Beds, *Pachygenelus* and *Trithelodon*. These Red Beds lie immediately below the Cave Sandstones. Gow (1980) is the most recent author to publish on *Pachygenelus* and *Trithelodon*. Little is known about these animals and Gow considers it not to be possible for there to be a squamosal-dentary contact in *Pachygenelus*. Until there are better descriptions of the material the term Ictidosauria should be used with great caution. It had a period of being a fashionable word, when it was used widely, even to include the tritylodonts. Despite the work of Crompton and Gow the original specimens are still poorly understood, and the term Ictidosauria is best confined to the limited sense in which it was first used by Broom.

## *Dromatherium* and *Microconodon*

Two enigmatic specimens will be included here, although they are clearly nothing to do with Broom's ictidosaurs. In 1857 two small lower jaws were discovered in a coal-mine, now long disused, at Egypt in North Carolina. They were called *Dromatherium* and *Microconodon*, and their age is somewhere near the boundary between Triassic and Jurassic. They were long considered mammals but Simpson (1926) classified them as cynodonts. His reasons for so doing were that the cheek-teeth lack fully divided roots, and, since the back end of the dentary is lacking in both specimens, it is impossible to be sure whether a dentary condyle is present or not. Cynodonts they certainly are not, and their systematic position is best regarded as quite uncertain. Kühne (1958) suggests the *Microconodon* may be a morganucodontid (Chapter 4).

## Teeth from the Württemberg Bone-bed

Two collections of isolated teeth may be mentioned here. One was made in the Rhaetic Bone-bed in Württemberg, W. Germany, and has been described by Erika von Huene (1933); and the other is from a deposit at Hallau in Canton Schaffhausen, Switzerland, and was described by Peyer (1956). They hint at a rich fauna of which we know practically nothing, and which we certainly do not understand. The material has been recently redescribed by Clemens (1980).

## Tritylodontidae

The tritylodonts* are a group of herbivorous cynodont derivatives. They were rather small herbivorous animals, varying in size from a rabbit to a beaver. The largest is *Tritylodontoideus maximus* from the South African Red Beds with a skull 230 mm long (Fourie, 1962). Tritylodonts are first

*See Kermack, D. M. (1982) for an up-to-date bibliography.

found in deposits of Upper Triassic age, and were the last major group of the mammal-like reptiles to appear (Tables 1.2 and 3.1). They continue until the Middle Jurassic, long after all the other mammal-like reptiles have disappeared. Although they are in no way ancestral to the mammals, they are an interesting group in their own right, and worthy of a brief description here.

The first skull of *Tritylodon* was picked up during military operations in Basutoland before 1884. It was described by Owen (1884), and at once acclaimed as the earliest mammal, since the cheek-teeth had fully divided roots, at that time a character only known in the Mammalia. The vital back end of the skull with the taxonomically all-important jaw articulation was missing. In 1943 Kühne published a preliminary description of *Oligokyphus* from Somerset and demonstrated there was no squamosal-dentary articulation; and so the tritylodonts were reptiles and not mammals after all.

Good material of another tritylodont, *Bienotherium*, was discovered in Yunan, south-west China by a team working under the direction of C.C. Young (1947). These specimens confirm the absence of the squamosal-dentary joint and the consequent reptilian nature of the tritylodonts. Although Fourie (1968) claims that there is a squamosal-dentary articulation in *Tritylodontoideus*, this observation needs confirmation in view of the condition of the jaw-joint in the other tritylodonts.

The tritylodonts are all very similar, and *Oligokyphus* from the Rhaetic Bone-bed of Germany and from the Lower Jurassic fissures of the Bristol Channel region will be used as an example of the group (Figures 3.2 and 3.3). The best material comes from Windsor Hill in Somerset, and was discovered and subsequently described by Kühne (1956). *Oligokyphus* was about the size of a rabbit, with a skull which superficially looks quite mammal-like. The post-orbital bone is lost, so that the orbit is confluent with the lateral temporal fossa behind, as it is in most mammals. The pineal foramen is closed and there is a fully developed bony secondary palate. The braincase and the palate are separated by a short girder formed by the pterygoid and parasphenoid: in the cynodonts this girder is long. These characters give a general mammalian appearance to the skull. There is a massive zygomatic arch to which the powerful masseter muscles for closing the jaw were attached.

The teeth of tritylodonts are, in some ways, the most interesting thing about them. There are no canine teeth, only incisors and cheek teeth, with a prominent diastema between them. In the upper cheek-teeth there are three rows of cusps, while in the lower jaw there are only two rows of cusps. Each cusp has a crescentic cutting edge facing forwards (mesially) in the upper teeth and backwards (distally) in the lower teeth. Each individual cusp in the post-canine teeth closely resembles the tooth of a rasp and functions in exactly the same way. In occlusion, the two lower rows of cusps fitted into the spaces between the three rows in the upper teeth, forming a self-aligning

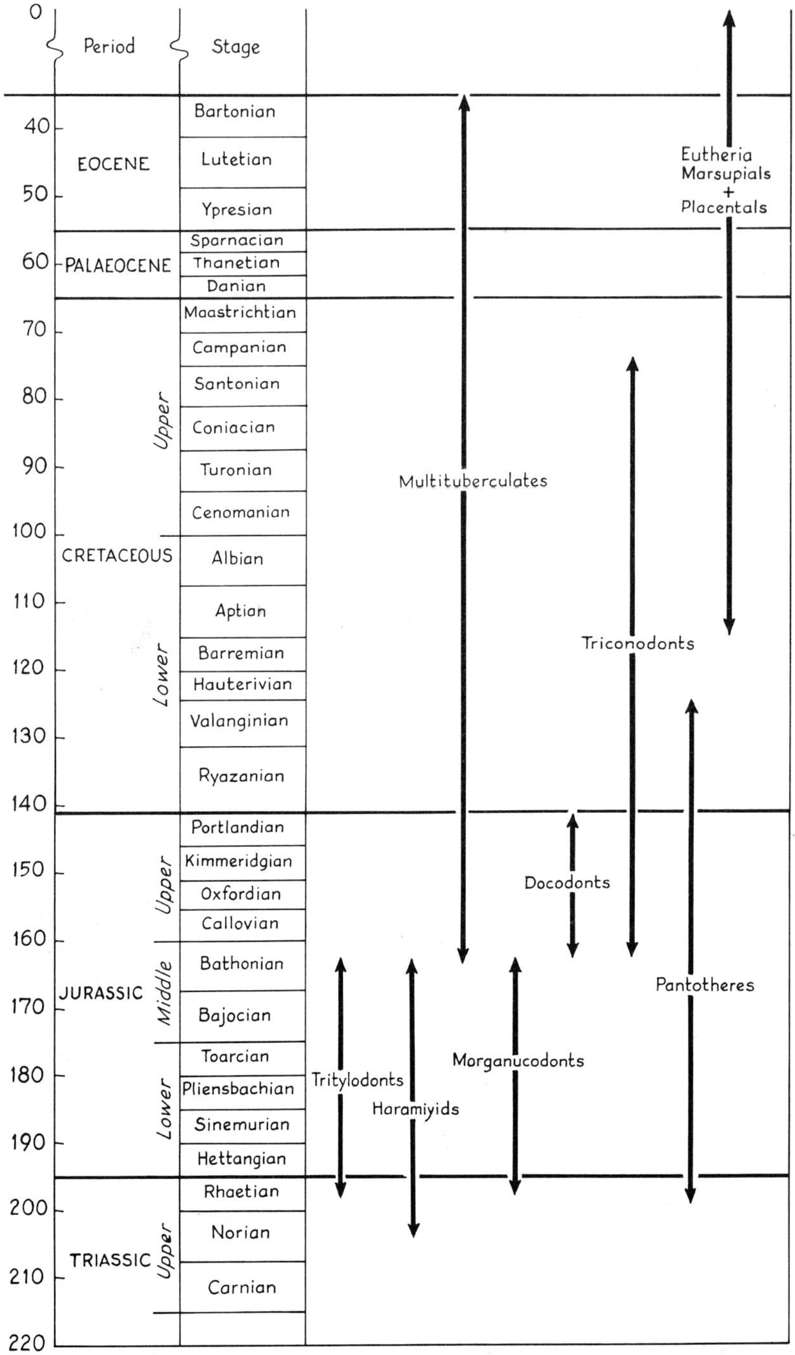


**Table 3.1:** The Range in Time of the Main Groups of Mesozoic Mammals

double rasp. This is perfectly symmetrical and so lacks any side-thrust. Since the primary function of the squamosal-dentary articulation is solely to resist such side thrust, it was not required and consequently did not evolve in the tritylodonts. The food was rapidly and efficiently shredded into small pieces by the rasping action of the post-canine teeth. For this action to take place it was necessary that the lower jaw had an anterio-posterior (propalinal) movement as the teeth came into occlusion.

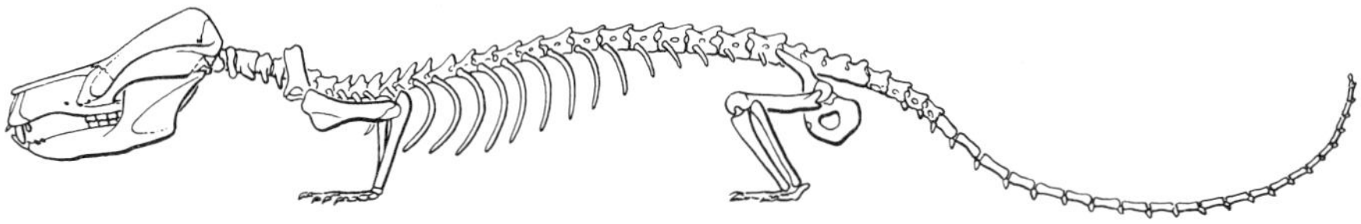

**Figure 3.2:** The Skeleton of *Oligokyphus*
Source: After Kühne, 1956

Kühne (1956) suggests that the propalinal movement in *Oligokyphus* was produced by the quadrate rocking on the squamosal. This we do not believe. Although some lateral rocking of the quadrate was necessary for sound conduction (Chapter 6), this was very limited in extent. Any considerable anterio-posterior movement of the quadrate in therapsids and early mammals would have induced highly undesirable movements of the footplate of the stapes in the fenestra ovalis. But there is no need at all to postulate this rocking. In herbivorous mammals propalinal movement of the jaw is

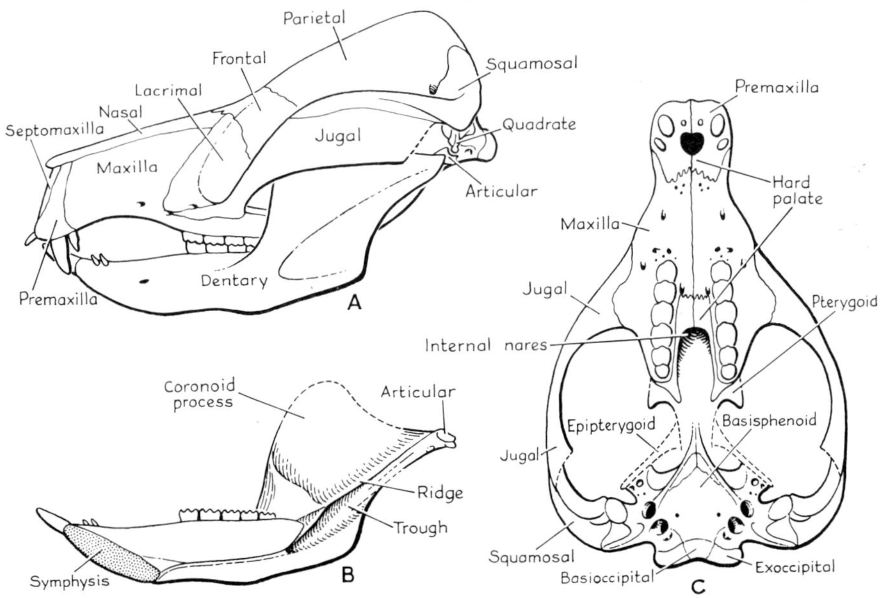


**Figure 3.3:** The Skull and Lower Jaw of *Oligokyphus*. A, lateral view of skull; B, medial view of lower jaw; C, ventral view of skull

produced by the jaw-articulation being placed high above the occlusal plane; and exactly the same condition obtained in the tritylodonts. In fact, in herbivorous tetrapods the jaw-articulation is always either well above or well below the occlusal plane for just this reason, while in carnivores the jaw articulation almost always lies in the occlusal plane. Observation of this fact is the easiest way of making an intelligent guess as to whether an animal is herbivorous or carnivorous.

The post-cranial skeleton of the tritylodonts is strikingly like that of the modern monotremes (Kühne, 1956). It is only more primitive in the lack of epiphyses and in the failure of the elements of the atlas vertebra to fuse together. Like monotremes the tritylodonts retain the coracoid and precoracoid bones in the pectoral girdle, with the precoracoid excluded from the glenoid. In the post-cranial skeleton, as in the skull with its double-rooted cheek-teeth, the tritylodonts are almost at the mammalian level of evolution. As nothing is known of their method of reproduction, only the lack of the squamosal-dentary articulation disqualifies them from being classified as mammals.

The tritylodonts continued on into the Middle Jurassic to become the last of the mammal-like reptiles to become extinct. They had no competition as all known Lower Jurassic mammals are carnivores or insectivores. The tritylodonts were doomed when the first herbivorous mammals, the multituberculates, appeared in the Upper Jurassic. It is not without interest that the Multituberculata parallel the tritylodonts sufficiently closely for *Tritylodon* itself to have been originally classified as a multituberculate.

## Haramiyidae

The Haramiyidae have been known for a long time, the first tooth was described in 1847; and not long after Charles Moore made a large collection from Holwell (Moore, 1867). Despite this they remain rather a mystery.

Haramiyids are known simply by isolated teeth. Figure 3.4 shows a tooth of *Thomasia antiqua* from Württemberg. No specimens of teeth in jaws have ever been found, and consequently no skeletal remains have ever been identified. These teeth come from the German Rhaetic Bone-bed, from Holwell quarry in Somerset, England, and from Hallau. The largest collection of haramiyid teeth now comes from the Rhaetic of France. The locality is at Saint-Nicolas-du-Port near Nancy in Lorraine (Sigogneau-Russell, 1978, 1983a, b) (see Chapter 4). A single tooth has recently been found in a fissure deposit in Pant Quarry, Glamorgan, South Wales, which has also yielded many specimens of early mammals (Kermack, 1975; Pacey, 1978). This last specimen is certainly Lower Jurassic in age, some of the others may be Triassic. The oldest haramiyid is a single tooth found during the preparation of a dinosaur from the Knollenmergel of Halberstat in W. Germany (Kühne,

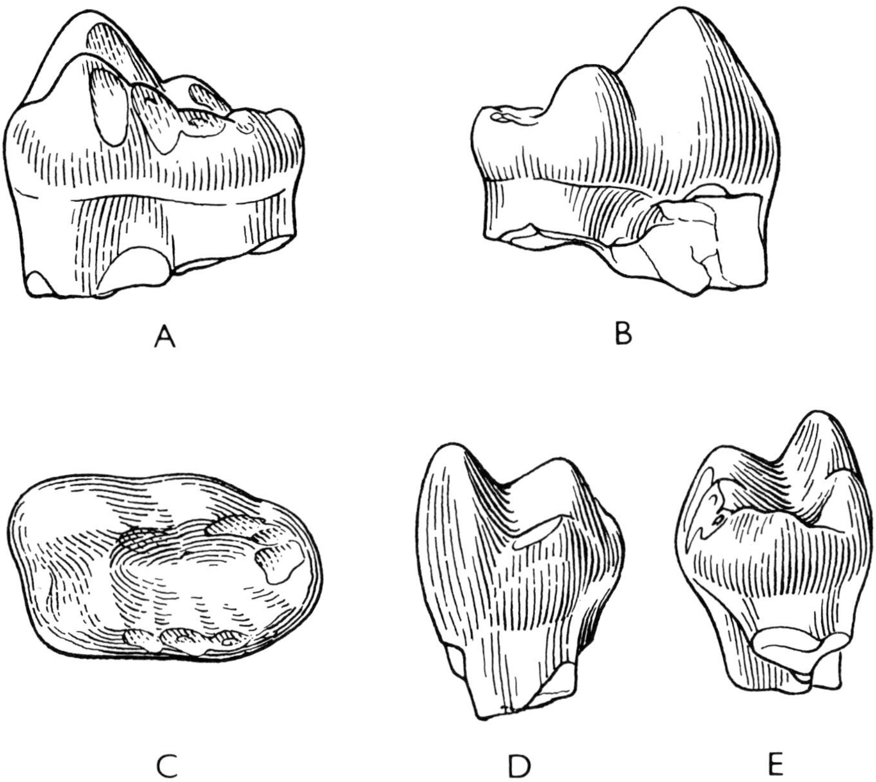


**Figure 3.4:** Different Views of a 'Molar' Tooth of *Thomasia antiqua*
Source: After Hahn, 1973

1968; Hahn, 1973). This is of pre-Rhaetian (Norian) age. If haramiyids are mammals it is the earliest mammalian fossil whose age is known with certainty (Table 3.1).

Not only are the haramiyids only known by isolated teeth, but only the cheek-teeth have been identified. The other teeth are quite unknown. It is possible, but not probable, that the two teeth from Holwell described by Parrington (1947) as *Eozostrodon* may be part of an haramiyid dentition. Haramiyid teeth are tiny; each has a pair of fully divided roots — this was why they were considered as mammals, like *Tritylodon* mentioned above; a row of two or three distinct cusps on one side of the crown and a row of three to six cusps on the other, with a trough between the two rows.

Professor G. Hahn, with his experience of the earliest multituberculates, considers the haramiyids to be ancestral to the former group (Hahn, 1973). The haramiyids are an enigmatic group, and will probably remain so until more is known of them beyond simply isolated cheek-teeth.

## References

Broom, R. (1932) *The Mammal-like Reptiles of South Africa, and the Origin of Mammals,* Witherby, London

Clemens, W. A. (1980) 'Rhaeto-Liassic Mammals from Switzerland and West Germany', *Abh. der Bayer. Staatssammlung für Paläont. und hist. Geol., 5*, 51–92

Crompton, A. W. (1958) 'The Cranial Morphology of a New Genus and Species of Ictidosaurian', *Proc. Zool. Soc. Lond., 130*, 183–216

Crompton, A. W. (1972) 'The Evolution of the Jaw Articulation in Cynodonts' in K. A. Joysey and T. S. Kemp (eds.), *Studies in Vertebrate Evolution,* Oliver and Boyd, Edinburgh

Fourie, S. (1962) 'Notes on a New Tritylodontid from the Cave Sandstone of South Africa', *Navoringe van die nasionale Mus., Bloemfontein, 2*, 7–19

Fourie, S. (1968) 'The Jaw Articulation of *Tritylodontoideus maximus', S. Afr. J. Sci., 64*, 255–65

Gow, C. E. (1980) 'The Dentitions of the Trithelodontidae (Therapsida: Cynodontia)', *Proc. R. Soc. Lond.* series B, *208*, 133–41

Hahn, G. (1973) 'Neue Zähne von Haramiyiden aus der deutschen Obertrias und ihre Beziehungen zu den Multituberculen', *Palaeontographica* A, *142*, 1–5

Huene, E. von (1933) 'Zur Kenntnis des Württembergischen Rhäbonebeds mit Zahnfunden neuer Säuger and säugerähnlicher Reptilien', *Ver. vaterländ. Naturkunde Württemberg Jh., 89*, 65–128

Heune, F. von (1942) *Die fossilen Reptilen des südamerikischen Gondwananlandes*, C.H. Beck, Munich

Kermack, D. M. (1982) 'A New Tritylodontid from the Kayenta Formation of Arizona', *Zool. J. Linn. Soc. Lond., 76*, 1–17

Kermack, K. A. (1975) 'The Complex of Early Mammals' in Problèmes actuels de paléontologie-évolution des vertébrés, *Colloques Internationaux du Centre National de la Recherche Scientifique, 218*, 563–72

Kermack, K. A., Mussett, F. & Rigney, H. W. (1981) 'The Skull of *Morganucodon', Zool. J. Linn. Soc., Lond., 71*, 1–158

Kühne, W. G. (1943) 'The Dentary of *Tritylodon* and the Systematic Position of the Tritylodontidae', *Ann. Mag. Nat. Hist.*, series 11, *10*, 589–601

Kühne, W. G. (1956) *The Liassic Therapsid* Oligokyphus, British Museum (Nat. Hist.), London

Kühne, W. G. (1958) 'Rhaetische Triconodonten aus Glamorgan, ihre Stellung zwischen den Klassen Reptilia und Mammalia und ihre Bedeutung für Reichart' sche Theorie', *Paläont. Zeit.,* series 5, *32*, 197–235

Kühne, W. G. (1968) 'Kimeridge (sic) Mammals and Their Bearing on the Phylogeny of the Mammalia', in Ellen T. Drake (ed.) *Evolution and Environment*, Yale University Press, New Haven

Moore, C. (1867) 'On the Abnormal Condition of Secondary Deposits', *Q. J. Geol. Soc. Lond., 23*, 449–568

Owen, R. (1884) 'On the Skull and Dentition of a Triassic Mammal (*Tritylodon longaevus*) from South Africa', *Q. J. Geol. Soc. Lond., 40*, 146–52

Pacey, D. E. (1978) *On a Tetrapod Assemblage from a Mesozoic Fissure Filling in South Wales*, Ph.D. thesis, University of London

Parrington, F. R. (1947) 'On a Collection of Rhaetic Mammalian Teeth', *Proc. Zool. Soc. Lond. 116*, 707–28

Peyer, B. (1956) 'Uber Zähne von Haramiyiden, von Triconodonten und von wahrscheinlich synapsiden Reptilian aus dem Rhät von Hallau, Kt. Schaffhausen, Schweiz', *Schweiz. paläont. Abh., 72*, 1–72

Romer, A. S. (1969) 'The Chañares (Argentina) Triassic Reptile Fauna. V. A New Chiniquodontid Cynodont, *Probelesodon lewisi* — Cynodont Ancestry', *Brevoria, 333*, 1–24

Romer, A. S. (1970) 'The Chañares (Argentina) Triassic Reptile Fauna. VI. A Chiniquodont Cynodont with an Incipient Squamosal-dentary Jaw Articulation', *Brevoria, 344*, 1–18

Romer, A. S. (1973) 'The Chañares (Argentina) Triassic Reptile Fauna. XVIII. *Probelesodon minor*, a New Species of Carnivorous Cynodont; Family Probainognathidae nov', *Brevoria, 401*, 1–4

Sigogneau-Russell, D. (1978) 'Découverte de Mammifères rhétiens (Trias supérieur) dans l'est de la France', *C. R. Acad. Sc. Paris* t. 287 série D: 991–3

Sigogneau-Russell, D. (1983a) 'A New Therian Mammal from the Rhaetic Locality of Saint-Nicolas-du-Port (France)', *Zool., J. Linn. Soc.*, *78*, 175–86

Sigogneau-Russell, D. (1983b) 'Nouveaux taxons de mammifères rhétians', *Acta palaeont. polonica, Warsaw*, *28*, 233–49, plates 10–12

Simpson, G. G. (1926) 'Mesozoic Mammalia. V. *Dromatherium* and *Microconodon*', *Amer. J. Sci.*, series 5, *12*, 87–108

Young, C. C. (1947) 'Mammal-like Reptiles from Lufeng, Yunnan, China', *Proc. Zool. Soc. Lond.*, *117*, 537–97

# 4 THE FIRST MAMMALS

## The Definition of a Mammal

The first mammals appear at the boundary between the Triassic and the Jurassic, most likely just before the end of the former period (see Table 3.1). Before describing them we must consider exactly what we mean by the term 'mammal'. Quite clearly no single character by itself expresses the concept of a mammal. To most people a mammal means a warm-blooded furry animal, whose mother nourished it from birth with her milk, a secretion of the modified sebaceous glands called mammae. Such a definition is of little value to a palaeontologist with only fossils to study. Skeletal characters are clearly needed, since, except in exceptional circumstances, nothing else remains in a fossil. In modern mammals there is a chain of three bones or auditory ossicles (the malleus, incus and stapes), which conducts sounds across the middle ear from the tympanic membrane to the fenestra ovalis of the inner ear. In all other tetrapods there is but a single auditory ossicle — the stapes. The mammalian jaw joint is between the condyle on the dentary and the glenoid on the squamosal, which is in contrast to the articular-quadrate joint of the Amphibia and reptiles. The reptilian articular and quadrate have become the malleus and incus respectively of the Mammalia, where the lower jaw consists of but a single bone, the dentary, as against the many-boned lower jaw of the other tetrapods. There are other skeletal characters, such as the three phalanges in the digits of a mammal, which had already been acquired by some of the later mammal-like reptiles. So the palaeontologist is left with a choice of just two characters, which are to some extent correlated with each other.

These characters are: (1) a squamosal-dentary jaw joint;
(2) a chain of three auditory ossicles.

The quadrate and the articular could not cease to form the jaw-hinge and pass into the middle ear to become the incus and malleus until after the squamosal-dentary joint had become established. The most satisfactory and logical choice would probably be to define a mammal on the basis of the presence of the three auditory ossicles. This definition has, however, a serious practical disadvantage. The auditory ossicles in mammals are tiny bones only loosely articulated with the skull. This is inevitable if these bones are to function as a device for transmitting sound across the middle ear. It is therefore not surprising that they are rarely preserved even in Tertiary mammals. *Morganucodon watsoni* is the only Mesozoic mammal in which traces of them have been found. *Morganucodon* is more completely known than any other Mesozoic mammal, and yet only a few incomplete stapes have been recovered. So, if the presence of the auditory ossicles is the criterion used for deciding if a specimen is a mammal, we are forced to rely on negative evidence. In other words, if there be no trace of the reptilian quadrate-articular jaw-joint, it can be assumed that the quadrate and articular have passed into the middle ear to become the incus and malleus. No animals are known which have a squamosal-dentary jaw-joint and only a stapes in the middle ear, the quadrate and articular having been lost. Whilst it is highly improbable that such an animal could have existed, we cannot be quite sure that it did not. Too much conviction of omniscience is unwise, also experience has shown that negative characters are treacherous. Much better to have a positive character so that it can be quite certain whether or not it is present. This is why most workers have accepted the first definition, namely, a mammal is a tetrapod with a squamosal-dentary jaw articulation.

Provided its limitations are clearly recognised, this is an excellent definition as the dentary is one of the most robust bones in the body and therefore most likely to be preserved. Indeed many species of Mesozoic mammals are known only by the lower jaw and even if the condylar area is not preserved a reasonable guess can be made about it from the condition of the cheek-teeth as we will see from Chapter 5. The limitations of the definition arise from the nature of the evolutionary process. Evolution is slow, continuous and imperceptible. The evolution of major taxa does not differ in any fundamental way from the evolution of one sub-species from another. Taxonomy is for the convenience of biologists and is in their minds and not in the universe of living things. The ability to classify depends upon breaks in the fossil record and if there were a complete and perfect record, taxonomy in its present form would be impossible. Taxonomic boundaries are placed in these breaks and that is why until the discovery of *Morganucodon* and its allies after the Second World War there was no problem in defining the Mammalia. Now the general consensus is to draw the boundary just below *Morganucodon* since its immediate ancestors are unknown.

## The Time of Origin of the Mammals

Two points must be stressed: firstly, the time of the origin of the Mammalia depends upon the character chosen (Table 4.1). If the attribute is the possession of the dentary-squamosal jaw-joint, then the mammals arose close to the boundary between the Triassic and the Jurassic. If the development of a fully mammalian ear with three auditory ossicles be chosen, then the first mammals probably arose in the Middle Jurassic. However, if the once popular definition of a mammal having a lower jaw composed of only one bone, the dentary, be used, then the first mammals are of Upper Jurassic age. There is a time span of some $30 \times 10^6$ years between the Triassic/Jurassic boundary and the beginning of the Upper Jurassic, and this

| Ma | Period | Series | Stage | Mammalian characters |
|---|---|---|---|---|
| | CRETACEOUS | Upper | | |
| | | Upper | Cenomanian | |
| 100 | | Lower | Albian | Fully-developed tribosphenic teeth. |
| | | Lower | Aptian | |
| | | Lower | Barremian | |
| | | Lower | Hauterivian | |
| | | Lower | Valanginian | Talonid of teeth with accessory cusps. |
| | | Lower | Ryazanian | |
| | JURASSIC | Upper | Portlandian | Lower jaw consists of only a single bone – the dentary. |
| | | Upper | Kimmeridgian | Loss of precoracoid & interclavicle; interclavicle & coracoid reduced & fused with scapular. |
| 150 | | Upper | Oxfordian | |
| | | Upper | Callovian | |
| | | Middle | Bathonian | Loss of quadrate/articular (reptilian) jaw articulation. Quadrate & articular concerned solely with hearing. |
| | | Middle | Bajocian | |
| | | Lower | Toarcian | |
| | | Lower | Pliensbachian | |
| | | Lower | Sinemurian | No evidence teeth replaced more than once. Tritubercular teeth. Cheek teeth come in definite shearing occlusion. |
| | | Lower | Hettangian | Squamosal/dentary jaw articulation. Bony external nares confluent. |
| | TRIASSIC | Upper | Rhaetian | Loss of prefrontal & postorbital bones from the skull. Precoracoid excluded from glenoid. |
| 200 | | Upper | Norian | Teeth with more than a single root. |
| | | Upper | Carnian | |
| | | Middle | Ladinian | |
| | | Middle | Anisian | |
| | | L'r | Scythian | Loss of postorbital bar. Bony 2ndary palate. Cheek teeth with cusps |
| | PERMIAN | Upper | Tartarian | Limited tooth replacement in upper canines. Passage for food in mouth functionally separated from airway. Phalangeal formula 2 3 3 3. Tympanum in lower jaw? |
| | | Upper | Kazanian | Teeth fully differentiated into incisors, canines & cheek teeth. |
| 250 | | Lower | Kungurian | |
| | | Lower | Artinskian | Beginning of differentiation of teeth into incisors, canines & cheek teeth. |
| | | Lower | Sakmarian | |

**Table 4.1:** Time of Known Appearance of Mammalian Characters

is equivalent to half of the Tertiary epoch. The difference in age is therefore considerable.

The second point is less obvious, but perhaps more significant. When we ask the question 'Are the Mammalia polyphyletic?' we must be careful to consider what is really meant. If we are saying 'Were the major taxa of the Mammalia already definitive and separate from each other when they crossed our chosen boundary?', then the answer is unequivocally 'Yes' on any of the three definitions given in the last paragraph. If we ask the quite different question 'Are the major taxonomic divisions of the Mammalia more closely related to each other than to any other taxonomic group?', then the answer has to be at the moment that we simply do not know. Having made our caveats, a mammal will be defined by the possession of a squamosal-dentary jaw articulation.

## Primitive Mammals

*Morganucodon* is about as primitive as a mammal can be and the anatomy of this once widespread genus is known in great detail as its remains are also

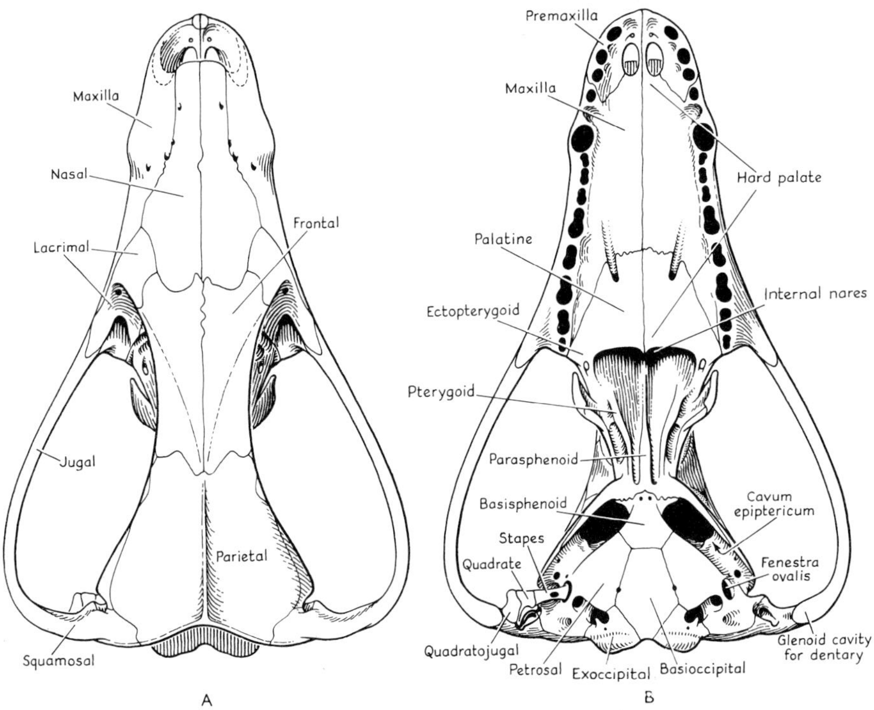


**Figure 4.1:** The Skull of *Morganucodon*. A, dorsal view; B, ventral view. See Figure 4.9 for lower jaw

abundant. They are found in the Welsh fissures (Kühne, 1958; Kermack, Mussett and Rigney, 1973), whilst those of two other species are found in the Lu Feng Beds of Yunan in south-western China (Rigney, 1963; Young, 1978). The age of the fissure deposits is not certain, but is probably Lower Jurassic (Sinemurian) and the Lu Feng Beds are also Lower Liassic or possibly slightly earlier, Rhaetian. A closely related form, *Erythrotherium* (Crompton, 1974), which may really be a member of the genus *Morganucodon*, is found in the Cave Sandstones of Lesotho. The age of these sandstones is uncertain; they may also be Lower Liassic, but Rhaetian seems most likely. Also related in some way to *Morganucodon* but how closely is not known, are *Eozostrodon* (Parrington, 1947) from England (probably Rhaetian) and *Megazostrodon* (Crompton, 1974) from the Lesotho Red Beds lying immediately below the Cave Sandstones. *Sinoconodon* from the Lu Feng Beds is rather enigmatic, but is probably related to the others. All are members of the sub-class of the Mammalia called the Atheria. The only living Atheria are the monotremes of Australia and New Guinea; all other living mammals are put into the other sub-class, the Theria. The first known Theria are found in the same Welsh fissures as *Morganucodon* but they will be discussed later.

Earlier mammals than *Morganucodon* are known from the Rhaetic of Europe, but these are isolated teeth. By far the most important collection comes from Saint-Nicolas-du-Port near Nancy in Lorraine (Russell, Russell and Wouters, 1976; Sigogneau-Russell, 1978, 1983a, b). This is probably the oldest mammalian fauna whose age is known with some confidence, and its full publication is awaited eagerly. The fauna includes morganucodonts, docodonts, kuehneotheriids and haramiyids.

Recently, Lower Jurassic mammals have been reported from India (Datta *et al.*, 1978; Datta, 1981) and North America (Jenkins, 1982). The Indian specimens came from the Kota formation in the Deccan, which is of Upper Liassic, or possibly Middle Jurassic, age (Datta, 1981). He describes in this paper a kuehneotheriid tooth — *Kotatherium haldani*. Jenkins mentions two morganucodont skulls and some other mammalian material from the Kayenta Formation of Gold Springs, Arizona.

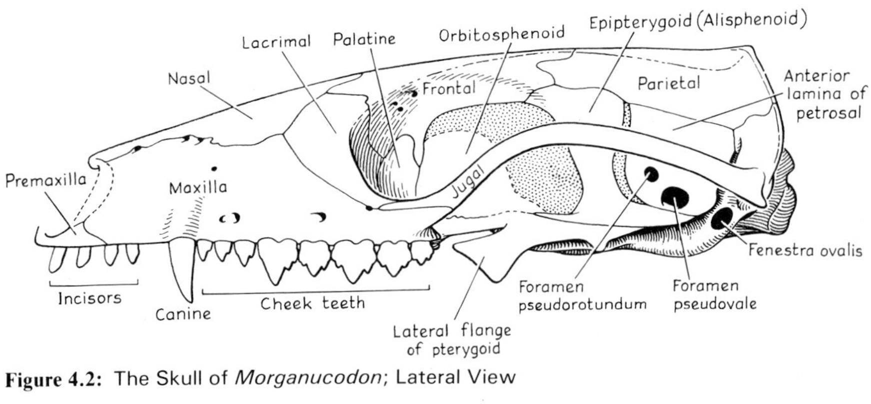


**Figure 4.2:** The Skull of *Morganucodon*; Lateral View

## *Morganucodon* — an Early Atherian Mammal

*Morganucodon* was a small animal about the size of the English mole. Figures 4.1 to 4.4 show the reconstructed skull, which looks more mammalian than does the skull of any mammal-like reptile, because of the relative proportions of the different parts. The braincase is relatively larger to the rest of the skull than it is in any mammal-like reptile. Part of this apparent increase in the size of the braincase is due to small vertebrates having, relative to the rest of their bodies, larger brains than their bigger relatives. Although most known mammal-like reptiles are considerably larger than *Morganucodon*, part of the increase in size of the latter's brain is real. Associated with the shape of the braincase is the shape of the palate. The

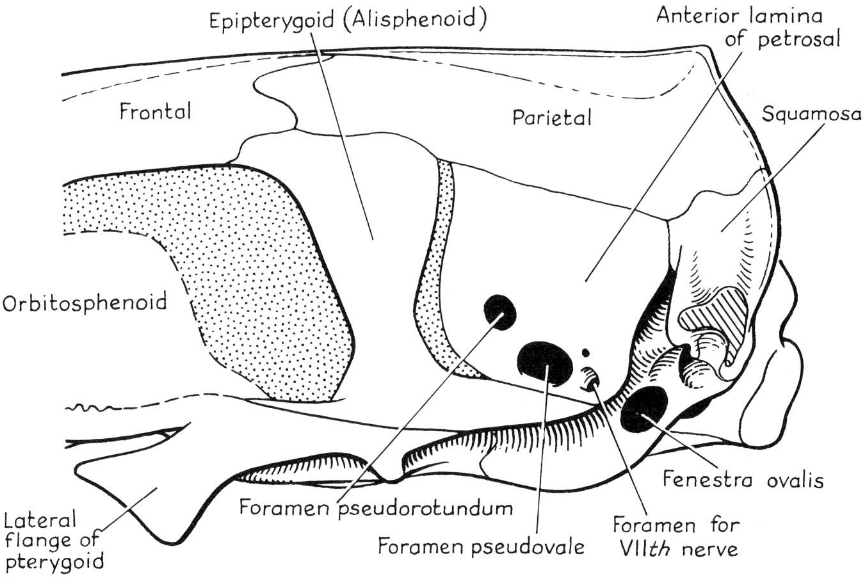


**Figure 4.3:** The Skull of *Morganucodon*; Lateral View of the Braincase with the Zygomatic Arch Removed

ventral view of the skull of a mammal-like reptile such as *Cynognathus* (Figures 4.5 to 4.8) shows a long gap between the back of the palate and the front of the braincase. This is bridged by a powerful girder formed by the parasphenoid rostrum, flanked by the quadrate processes of the epipterygoid and pterygoid. In modern mammals and *Morganucodon* there is no gap and so the latter's skull resembles that of modern mammals rather than that of the cynodonts. Other characters contributing to the overall mammalian appearance of the skull of *Morganucodon* are: the unpaired external nares, the loss of the prefrontal bone, the disappearance of the post-orbital bar and with it the post-orbital bone, the closure of the pineal foramen and the large orbitosphenoid forming the median wall of the orbit.

The dentition of *Morganucodon* looks, and is, fully mammalian with incisors, canines, premolars and molars (Figures 4.2 and 4.9). All but the molars are replaced, probably once as they are in placental mammals, but not incidentally in marsupials. The mammalian condition of the dentition may or

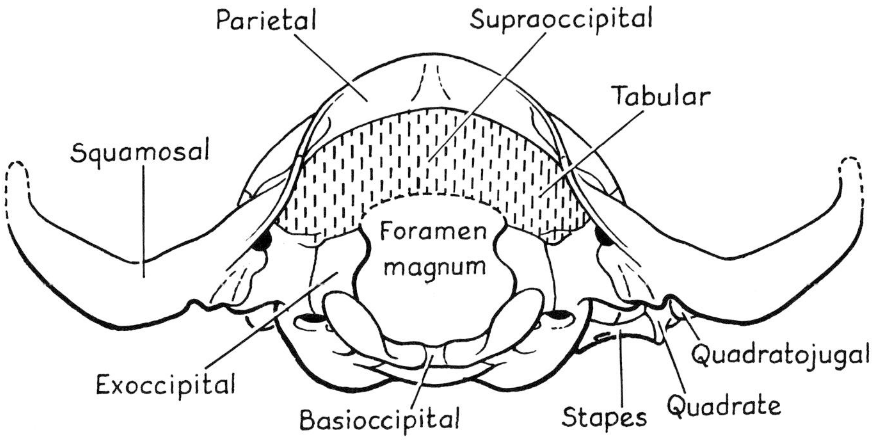


**Figure 4.4:** The Skull of *Morganucodon*; Occipital View

may not be a primitive mammalian attribute, so the resemblance here between *Morganucodon* and modern placentals may be quite fortuitous. However, *Morganucodon* used its teeth in a thoroughly mammalian way as we have shown in Chapter 5 and this is reflected in their appearance.

However, when the skull is looked at more closely, there are features about it that seem strange, even omitting the double jaw articulation and that the lower jaw consists of more than one bone, the dentary. Again one of these is a matter of proportions. The jaw articulation is at the posterior end of the skull in the same vertical plane as the occipital condyles, which is typical of mammal-like reptiles and not found in any Recent mammal. Another difference, which is not simply a difference in proportion, is that the lateral wall of the braincase is largely formed by an anterior sheet-like extension of the petrosal bone which houses the inner ear. This sheet of bone is called the 'anterior lamina' of the petrosal and in front and slightly lateral to it is a rather narrow plate of bone, the alisphenoid (epipterygoid) (Figure 4.3). The two bones do not come into contact with each other, there being a small space between them (the cavum epiptericum). In the vast majority of living mammals, the marsupials and placentals, the structure of the lateral wall of the braincase is fundamentally different and formed by the squamosal and alisphenoid. In these animals two of the three main trunks of the Vth (trigeminal) cranial nerve leave the skull by means of foramina in the

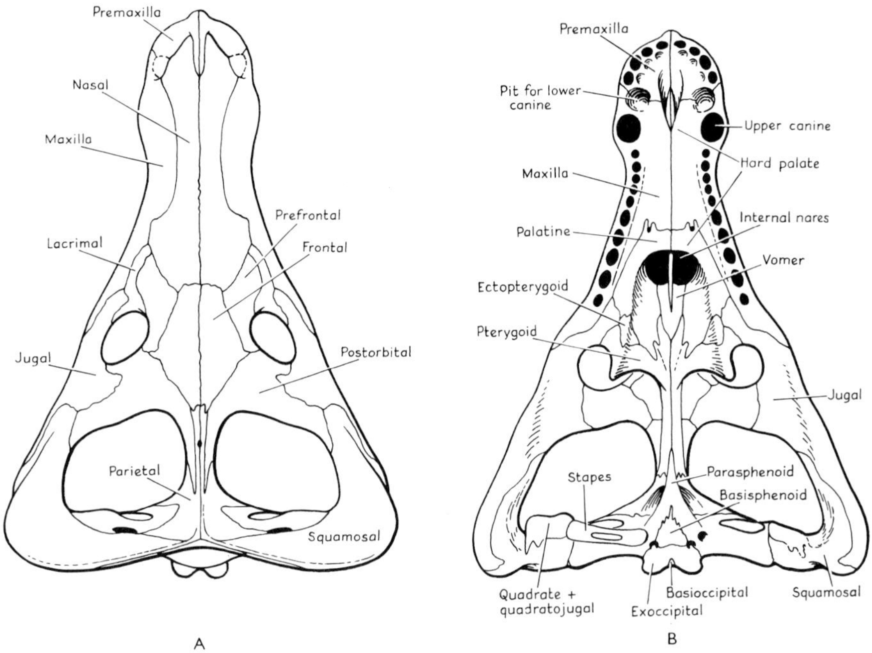


**Figure 4.5:** The Skull of *Cynognathus*. A, dorsal view; B, ventral view. See Figure 5.10 for lower jaw
Source: After Broili and Schröder, 1934

alisphenoid. The most posterior, the mandibular trunk, leaves via the foramen ovale, whereas the more anterior maxillary trunk passes out through the foramen rotundum. In *Morganucodon* (Figure 4.3) both trunks exit through the anterior lamina, the corresponding foramina being called the foramen pseudovale and the foramen pseudorotundum to distinguish them from those in the alisphenoid. Another feature is that the squamosal plays little part in the formation of the braincase.

Although *Morganucodon* differs radically from the living Theria in these particulars, it has similarities with living monotremes (Kermack and Kielan-Jaworowska, 1971). The latter have a large anterior lamina and tiny alisphenoid and the sharing of this specialised character gives the most powerful reason for placing both *Morganucodon* and the monotremes in the sub-class Atheria. They also have other characters in common: the presence of an ectopterygoid, a cochlea which is only slightly curved and no separate aqueductus cochleae, to name but three. These are primitive characters and carry much less weight in determining relationships. An advanced character possessed by cynodonts is the broad plate-like epipterygoid but in *Morganucodon* this bone is quite unspecialised, resembling closely that of a

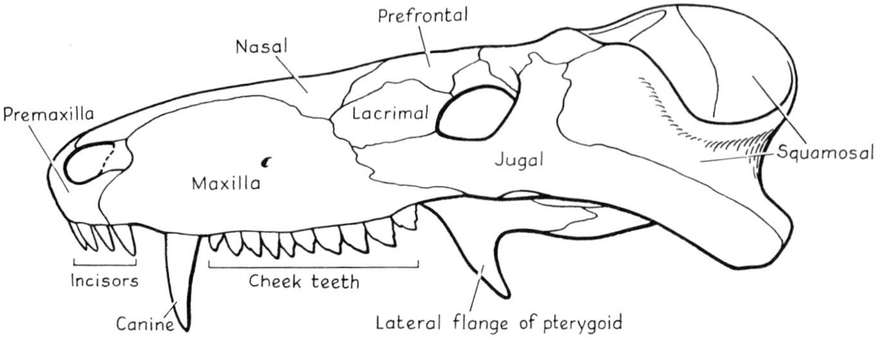


**Figure 4.6:** Lateral View of the Skull of *Cynognathus*
Source: After Broili and Schröder, 1934

Permian gorgonopsid. The epipterygoid has not been described in any other member of the Triconodonta but in the Multituberculata, another group of atherians, this bone has also become reduced. Thus in the Cynodontia and in the Theria the evolutionary trend is for the alisphenoid to become a broad sheet of bone, whilst in the Atheria the specialisation is for the epipterygoid (alisphenoid) to become reduced. In other words the advanced or apomorphic character in the Atheria is the reduced alisphenoid, whereas in

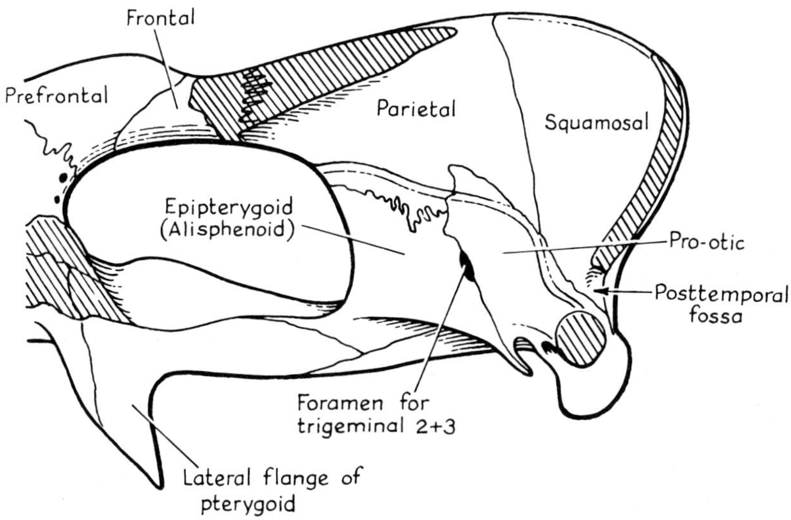


**Figure 4.7:** Lateral View of the Braincase of *Cynognathus* with the Zygomatic Arch Removed
Source: After Broili and Schröder, 1934

the Cynodontia and the Theria it is the broadened plate-like epipterygoid. So the Atheria on one hand and the cynodont-therian assemblage on the other form a pair of Hennigian 'sister-groups'. Unlike most workers in the field, we doubt if the Atheria evolved from the cynodonts. The latter may have given rise to the Theria but there is no certainty.

It may be argued that we are putting forward a single character classification, which is usually unwise, but there is other supporting evidence. In *Morganucodon* one of the main veins of the head, the *vena capita lateralis*, passes below the lateral flange of the petrosal (Figure 4.3; Kermack *et al.*, 1981). According to Watson (1911), the same vein passes *above* the lateral flange in the cynodonts. This is again a fundamental difference, since it is impossible to derive one condition from the other.

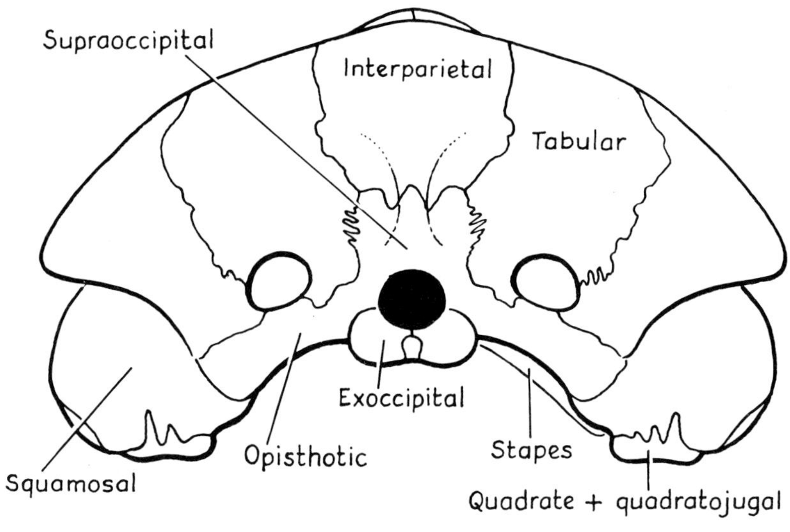


**Figure 4.8:** Occipital View of the Skull of *Cynognathus*
Source: After Broili and Schröder, 1934

Figure 4.9 shows the lower jaw of *Morganucodon* reconstructed with the accessory jaw bones in place. The reptilian (quadrate-articular) and the mammalian (squamosal-dentary) jaw articulations can be seen lying side by side.

A second and independent character showing the Theria and the Atheria as sister groups comes from a consideration of their teeth. In the former the molar teeth are built on a basically triangular arrangement of the cusps. In other words, the molars are derivable from what is known as the trituberculo-sectorial tooth (see Chapter 5), or are of a form which is clearly ancestral to that pattern of tooth. The unequivocal appearance of the therian molar

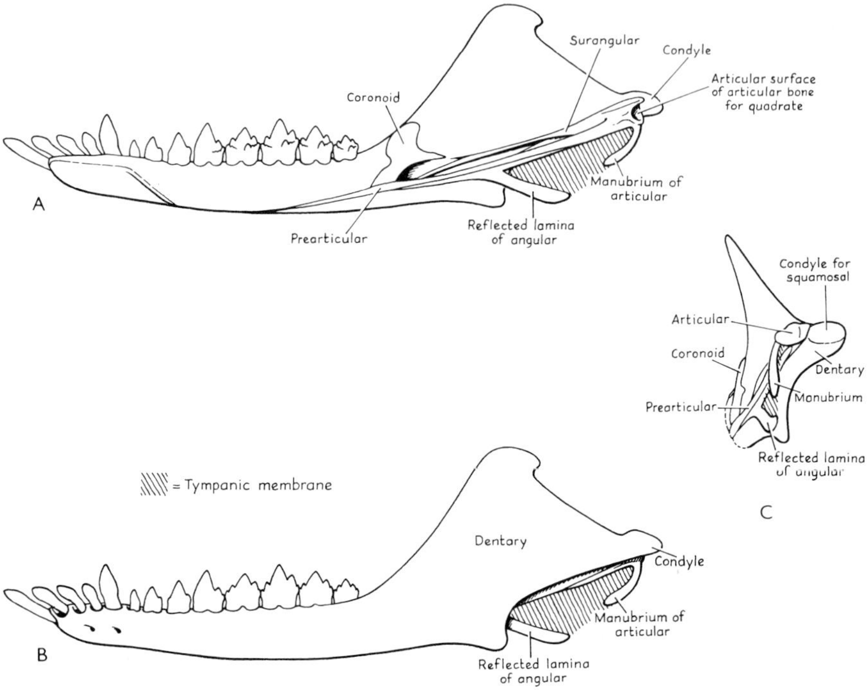


**Figure 4.9:** Reconstruction of the Lower Jaw of *Morganucodon*, with the Position of the Tympanic Membrane Indicated. See Figure 5.10 for the lower jaw of *Cynognathus*

pattern in *Kuehneotherium*, for long the earliest known therian mammal and coeval with *Morganucodon*, is one of the more remarkable discoveries of the last decades.* The terminology based on cusp homologies worked out from the modern Theria can be applied without much difficulty to the molars of *Kuehneotherium*. In two of the groups of the Atheria, the multituberculates and the docodonts, the teeth become complex but the basic triangular pattern of the cusps, so characteristic of the Theria, is always absent, and the terms used to describe cusp homologies in the latter cannot be applied to them. This is true even though in some docodonts the molar teeth function in a way which parallels closely the trituberculo-sectorial molars of the therian mammals. Thus the terms used in describing therian cusp homologies cannot be applied to them.

The post-cranial skeleton of *Morganucodon* is similar to that of both the contemporary tritylodonts and the modern monotremes. There is nothing

*Recently Sigogneau-Russell (1983a, b) has described kuehneotheriid-like teeth from Saint-Nicolas-du-Port. These are now the earliest known therian teeth.

distinctively mammalian about it, and the Mammalia cannot be defined from the post-cranial skeleton. The skeleton of *Morganucodon* is simply at the level to be expected in a member of the therapsid-mammalian stock early in the Jurassic. It shows an advance upon that of the cynodonts in the exclusion of the precoracoid from playing any part in the formation of the glenoid for the articulation of the head of the humerus. Neither *Morganucodon* nor the tritylodonts (Kühne, 1956) show growth by epiphyses; monotremes on the other hand do show this pattern of growth, which they must have acquired independently of the Theria, which also show it. There is no difficulty here: epiphyseal growth is also found in some living lizards, which have also acquired it independently.

The other early Atheria, all of them probably of Lower Jurassic or possible Upper Triassic age, may be grouped with *Morganucodon* in the Order Morganucodonta. Two of them, *Erythrotherium* and *Megazostrodon*, are known by reasonably complete skulls (Crompton, 1974). When full descriptions of them have been published it will be possible to have a better understanding of them.

### *Kuehneotherium* — an Early Therian Mammal

*Kuehneotherium*, the earliest known therian mammal occurs along with *Morganucodon* in the fissures of South Wales (Kermack, D. M. *et al.*, 1968). Unfortunately it is much less well known as the material is limited to teeth and almost invariably toothless jaw fragments. The reason why the jaws of *Morganucodon* are found containing teeth whereas the teeth of *Kuehneotherium* are seldom found *in situ* in the jaw is not without interest. When the cheek-teeth of *Morganucodon* have finished erupting the apices of the roots expand to give an 'elephant's foot' appearance. Roots of this pattern are described as taurodont and lock the teeth firmly in their sockets, so the teeth do not drop out when all the soft parts of the jaw have decayed. This is important in a fissure deposit, where the animals decayed completely before the bones were transported by water to their final resting place, being broken and rolled in the process. Not all morganucodonts have taurodont roots — *Eozostrodon* and *Megazostrodon* lack them; and they seem not to have been present in early Theria. The teeth of *Kuehneotherium* lack taurodont teeth so that they always fall out of the jaw, which is a pity. Assembling isolated teeth into a dentition is a difficult and unsatisfactory process and had *Kuehneotherium* had taurodont roots a lot more would be known about it.

*Kuehneotherium* was certainly an insectivore with a dentition well-suited to cutting up the tough chitinous exoskeletons of its prey. Its molar teeth are of the basic therian pattern showing the typical arrangement of the main

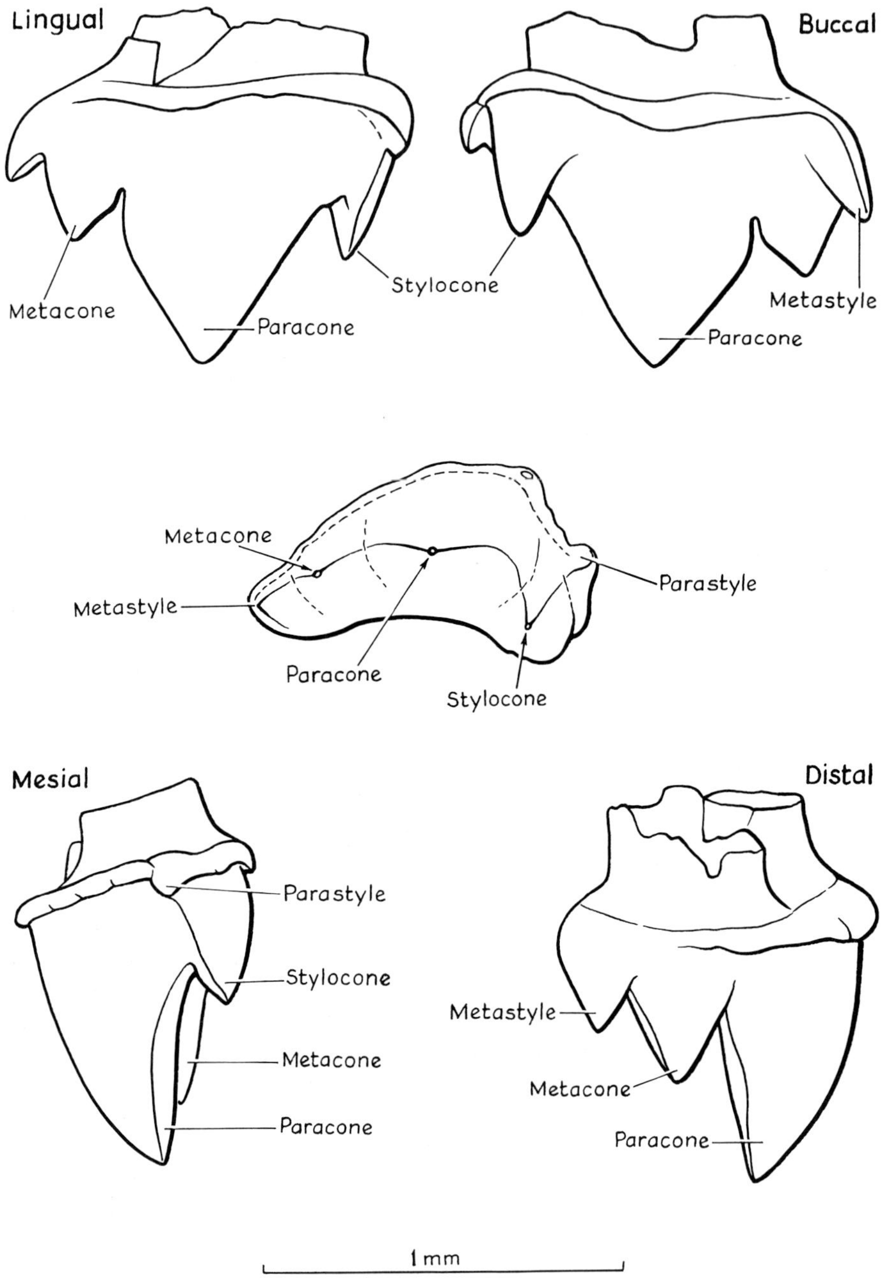


**Figure 4.10:** Upper Molariform Tooth of *Kuehneotherium*
Source: After Kermack, Kermack and Mussett, 1968

cusps in a triangle. In occlusal view* the outline of the upper molars is roughly a right-angled triangle with its apex directed lingually. The main cusp or *paracone* is at the apex of the triangle and the cutting edge or *crest* runs disto-buccally along the hypotenuse. This crest bears a small cusp, the *metacone*, and finally terminates in the *metastyle*, a small cusp lying slightly buccally to the metacone. The *cingulum* or narrow shelf runs around the whole tooth terminating distally in the metastyle and bearing a pair of small *accessory cusps* mesially. Finally, the *stylocone*, a large cusp comparable in size to the metacone, arises mesially from the cingulum and buccally to the paracone (Figure 4.10).

The lower molars are also triangular in occlusal view (Figure 4.11), but the apex is directed buccally and not lingually as in the upper teeth. Again the main cusp, the *protoconid*, is situated at the apex with crests running from it mesio-lingually and disto-lingually. Each crest ends in a small cusp: the mesial cusp is called the *paraconid* and the distal one the *metaconid*. These three main cusps form the *trigonid*. The cingulum is largely confined to the lingual side of the tooth in the lower molars, but laps around on the buccal side at the mesial end. Mesially the cingulum bears a pair of accessory cuspules similar to those of the upper teeth, while distally it ends in a single small cusp called the *hypoconulid*. The latter and the tiny distal extension of the cingulum adjacent to it are called the *talonid*, which becomes much larger and more important in later Theria. Thus the basic pattern of therian molar teeth is already established early in the Jurassic.

This triangular shape of the molar teeth is peculiar to the Theria and although in the Jurassic some of the Atheria have more complex teeth than any of the contemporary Theria, the cusps are always arranged in rows, never in triangles. The triangular shape of the molar teeth of *Kuehneotherium* is of great advantage to an insectivore. The shearing is done by the three main cusps in the lower molars moving against the three in the uppers, trigon shearing against the trigonid. Tough pieces of cuticle jammed between the molars will try to push the lower jaw laterally and so force the teeth out of occlusion, but they will also push the lower jaw longitudinally and thus neighbouring teeth into closer occlusion. To protect the gums from damage, the metastyle in the upper teeth interdigitates between the mesial accessory cuspules, thus preventing food from being impacted between the teeth. The hypoconulid and the pair of mesial cuspules behave in an exactly similar way in the lower molars (Figure 4.12). Lingually the cingulum in the upper molars expands in two places to form a shelf. The more mesial of these lies between the stylocone and the paracone and its function is to stop food being impacted between the mesial root and the gum by the action of the protoconid. The second and rather smaller shelf lies between the paracone

*See Figure 5.1 for details of the nomenclature used in describing the orientation of the teeth in the jaw.

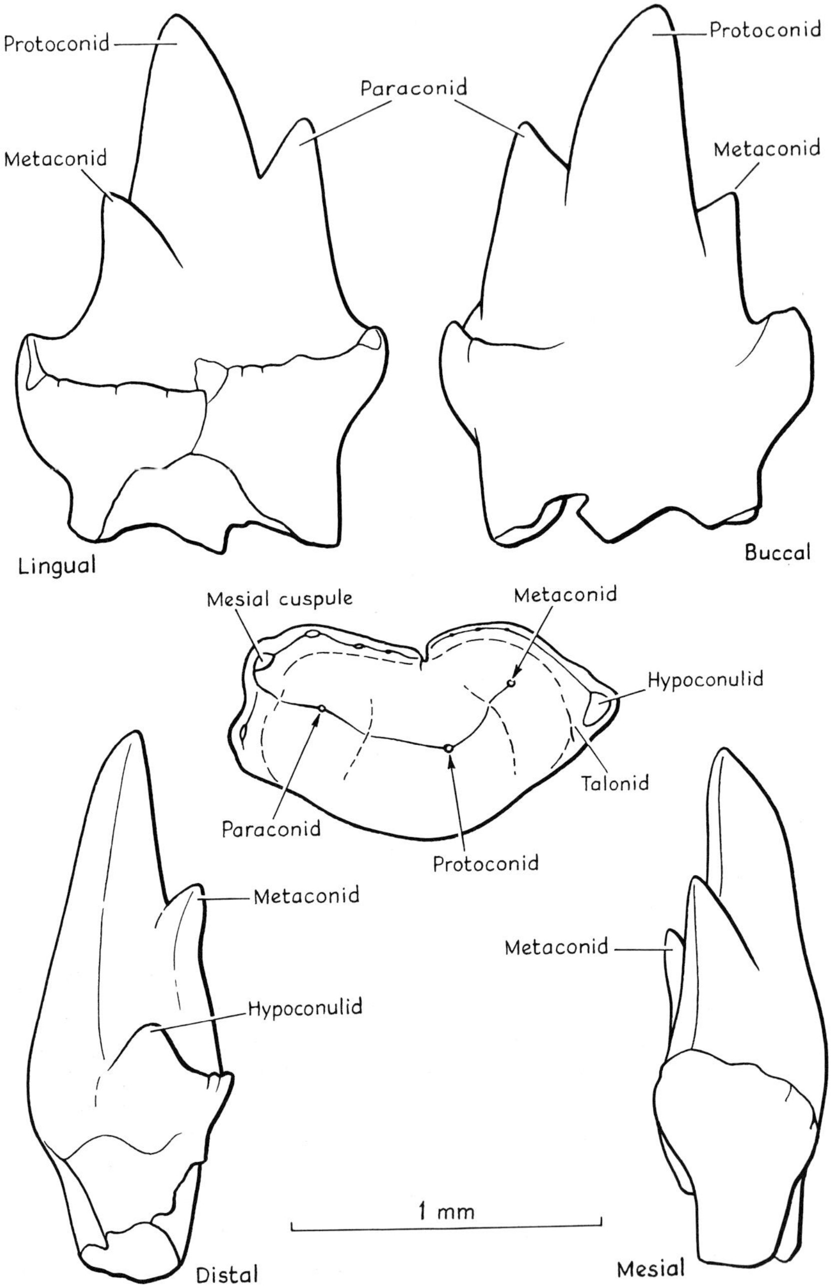


**Figure 4.11:** Lower Molariform Tooth of *Kuehneotherium*
Source: After Kermack, Kermack and Mussett, 1968

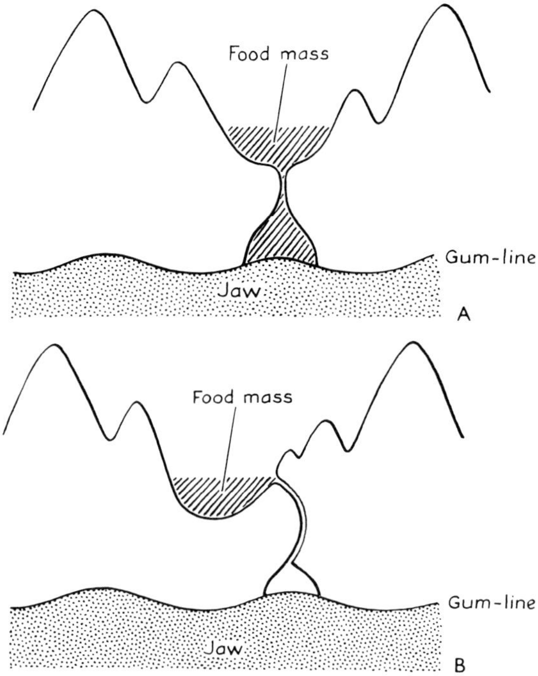


**Figure 4.12:** To Show How the Interlocking of Adjacent Teeth Protects the Gum from Damage
Source: After Kermack, Lees and Mussett, 1965

and the metacone; it protects the distal root of the upper molar from the impaction of food by the action of the paraconid. The molar teeth of all succeeding Theria can be derived without difficulty from the basic pattern exemplified by the molars of *Kuehneotherium*.

*Kuehneotherium* had a definite condyle on its dentary (Figure 4.13) and a large trough on the medial side exactly analogous to that in the lower jaw of *Morganucodon*. So there is no doubt the *Kuehneotherium* had a full complement of reptilian jaw bones including the quadrate-articular reptilian jaw-hinge as well as its mammalian joint between the squamosal and dentary.

Thus by the early Jurassic, the Atheria as represented by *Morganucodon* and its allies and the Theria as represented by *Kuehneotherium* had already become separated from each other as the two major divisions of the class Mammalia. Modern Atheria as represented by the monotremes of Australasia and the modern Theria represented by all other living mammals have lost their reptilian jaw-joint. Some of the reptilian jaw bones have been lost but some have been retained and changed their function. The quadrate and articular have become part of the chain of three bones which transmit sound from the ear-drum to the inner ear in modern mammals, although they

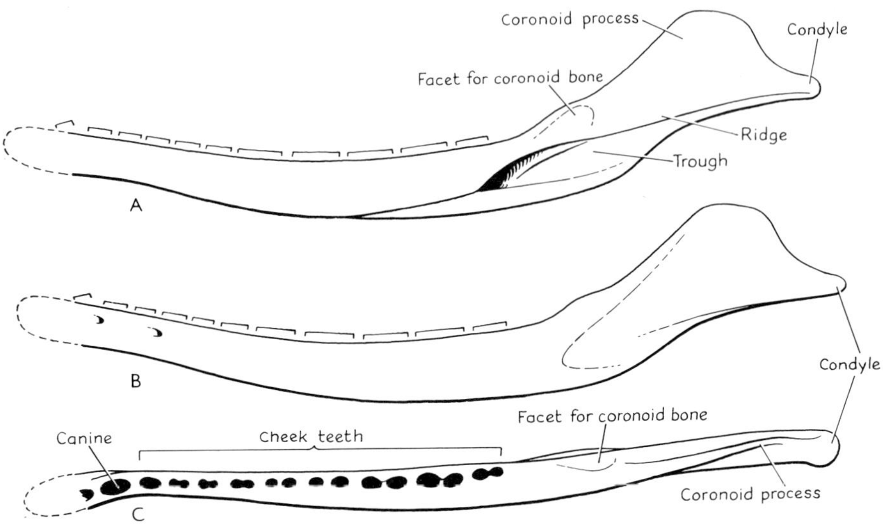


**Figure 4.13:** Lower Jaw of *Kuehneotherium*. Compare with Figure 4.9 — the lower jaw of *Morganucodon*

are now called the incus and malleus respectively. The angular has also left the lower jaw and is called the tympanic bone. How and why these changes came about will be explained in Chapter 6. Since the Theria and the Atheria separated from each other before changes in the middle ear had taken place, these two major groups must have evolved mammalian auditory ossicles independently. This is a most surprising fact.

## References

Broili, F. & Schröder, J. (1934) 'Zur Osteologie des Kopfes von *Cynognathus', Sitzungsber. Bay. Akad. Wiss. Math. Naturwiss., 95*, 95–128

Crompton, A. W. (1974) 'The Dentitions and Relationships of the Southern African Triassic Mammals, *Erythrotherium parringtoni* and *Megazostrodon rudnerae', Bull. Brit. Mus. (nat. hist.), 24*, 397–437

Datta, P. M. (1981) 'The First Jurassic Mammal from India', *Zool. J. Linn. Soc. Lond., 73*, 307–12

Datta, P. M., Yadagiri, P. & Rao, B. R. J. (1978) 'Discovery of Early Jurassic Micromammals from Upper Gondwana Sequence of Pranhita Godavari Valley, India', *J. Geol. Soc. India, 19*, 64–8

Jenkins, F. A. (1982) *News Bulletin, 126*, 74, Society of Vertebrate Palaentology, Gainsville, Florida, USA

Kermack, D. M., Kermack, K. A. & Mussett, F. (1968) 'The Welsh Pantothere *Kuehneotherium praecursoris', J. Linn. Soc. (Zool.), 47*, 407–23

Kermack, K. A. & Kielan-Jaworowska, Z. (1971) 'Therian and Non-therian mammals' in Kermack D. M. and Kermack K. A. (eds.), *Early Mammals, Zool. J. Linn. Soc., Lond., 50*, Suppl.1, 103–15

Kermack, K. A., Lees, P. M. & Mussett, F. (1965) '*Aegialodon dawsoni*, a New Tuberculo-sectorial Tooth from the Lower Wealden', *Proc. Royal Soc. B., 162, 535–54*

Kermack, K. A., Mussett, F. & Rigney, H. W. (1973) 'The Lower Jaw of *Morganucodon', Zool. J. Linn. Soc., Lond., 53*, 87–175
Kermack, K. A., Mussett, F. & Rigney, H. W. (1981) 'The Skull of *Morganucodon', Zool. J. Linn. Soc., Lond., 71*, 1–158
Kühne, W. G. (1956) *The Liassic Therapsid Oligokyphus*, Br. Mus. (nat. hist), London
Kühne, W. G. (1958) 'Rhaetische Triconodonten aus Glamorgan, ihre Stellung zwischen den Klassen Reptilia und Mammalia und ihre Bedeutung für Reichart'sche Theorie', *Paläont. Zeit.*, series 5, *32*, 197–235
Parrington, F. R. (1947) 'On a Collection of Rhaetic Mammalian Teeth', *Proc. Zool. Soc. Lond., 116*, 707–28
Rigney, H. W. (1963) 'A Specimen of *Morganucodon* from Yunnan', *Nature, 197*, 1122–3
Russell, D., Sigogneau-Russell, D. & Wouters, G. (1976) 'Une Dent d'aspect Mammalien en Provenance du Rhétien Français', *Géobios, 4*, 377–92
Sigogneau-Russell, D. (1978) 'Découverte de Mammifères Rhétiens (Trias Supérieur) dans l'est de la France', *C. R. Acad. Sc. Paris* t. 287 série D, 991–3
Sigogneau-Russell, D. (1983a) 'A New Therian Mammal from the Rhaetic Locality of Saint-Nicolas-de Port (France)', *Zool. J. Linn. Soc., Lond., 78*, 175–86
Sigogneau-Russell, D. (1983b) 'Nouveaux taxons de Mammifères rhetiens', *Acta palaeont. polonica, Warsaw, 28*, 233–49, plates 10–12
Watson, D. M. S. (1911) 'The Skull of *Diademodon*, with Notes on Those of Some Other Cynodonts', *Ann. Mag. Nat. Hist.*, ser. 8, *8*, 293–330
Young, C. C. (1978) 'New Materials of *Eozostrodon', Vertebrata palasiatica, 16*, 1–3

# 5 DENTITIONS, TOOTH-REPLACEMENT AND JAW ARTICULATION

Reptiles evolved into mammals under two separate pressures. The first was to improve hearing, so that they could become efficient nocturnal animals. This will be discussed in a later chapter (Chapter 6). The second pressure was dental: the unremitting drive to improve the efficiency of the teeth, initially for seizing and killing prey and later for cutting it up into conveniently sized pieces for swallowing. It will be noted that throughout we shall be talking about carnivores, as plant-eaters have such specialised dentitions that they have lost the capacity for further major advances. The future lay with the flesh-eater, and in particular with the insectivore with its relatively generalised dentition.

## Dental Terminology

When discussing teeth, the standard dental terms to orient them have to be used if ambiguity is to be avoided. Unfortunately the unfamiliarity of the terms used makes the story of the evolution of mammalian dentitions appear much more obscure than it really is. When the dentist looks into your mouth he will describe the vertical surface of the crown of the tooth facing the symphysis of the two sides of the lower jaw as the *mesial* surface (Figure 5.1). The opposite surface which faces towards the jaw articulation, he will call the *distal* surface. The vertical surface at right-angles to this which faces the tongue is the *lingual* surface. The opposite surface, which faces outwards towards the cheeks, is the *buccal* surface. The horizontal surface of the tooth, which meets the similar surface of the opposing tooth when the jaws close, is the *occlusal* surface. It is at right-angles to all the previously mentioned four faces. Finally the tip of the root is the *apex* of the tooth, with the gum-line marking the junction between root and crown.

A tooth is said to be *mesial* to another if it lies nearer to the symphysis of the jaw, and *distal* if it lies closer to the jaw articulation.

## Tooth Replacement

In primitive reptiles the marginal teeth, those on the premaxilla, maxilla and dentary, are all about the same size. Each tooth is implanted in a socket or *alveolus*, which is occupied by a succession of teeth, called the *tooth-family*

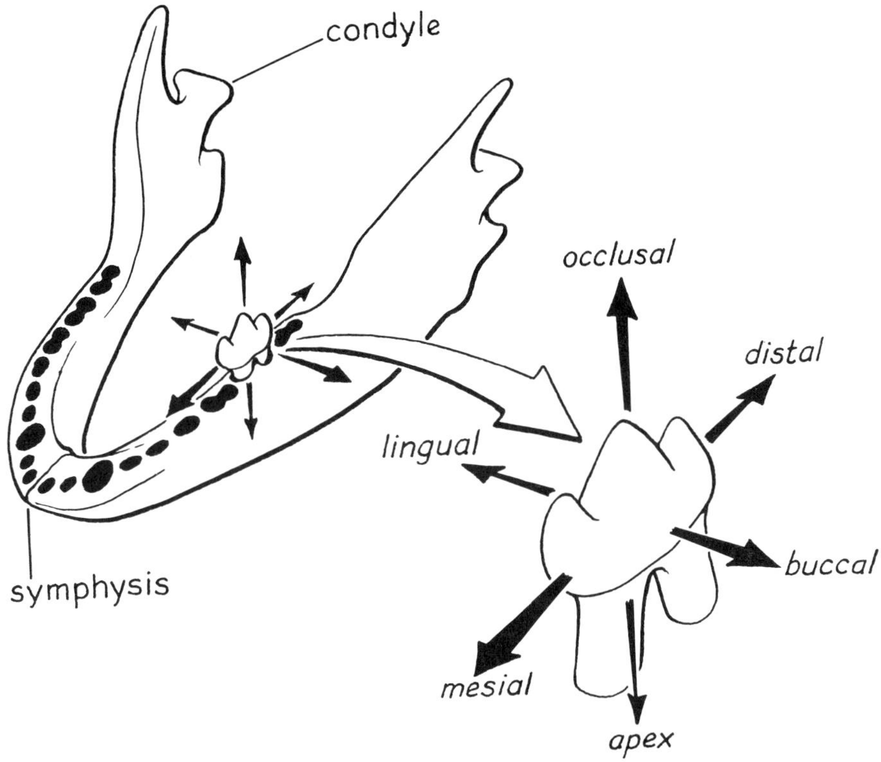


**Figure 5.1:** The Standard Dental Terminology for Orienting Teeth

of the alveolus. Each tooth arises from the dental lamina lingually, and moves buccally as it grows larger. The eldest member of a tooth family always forms the functional tooth. When a tooth has become worn out it is shed and replaced by the next member of its family. The latter moves into the buccal position, and becomes implanted in the alveolus as the functional tooth. This steady replacement of teeth goes on throughout the whole life of the animal.

The shedding of the tooth and its eventual replacement takes some considerable time, and to avoid embarrassing gaps in the dentition adjacent alveoli do not replace their teeth simultaneously. In consequence, there tends to be some synchronisation of the replacement of the teeth in alternate alveoli. So often the tooth row of a primitive reptile will show a pattern of functional and replacement teeth alternating (Figure 5.2). This is the rationale of alternate tooth replacement, which has, at times, been given an almost mystical significance. The term *distichial replacement* is sometimes applied to it (Parrington, 1936; Kermack, 1956).

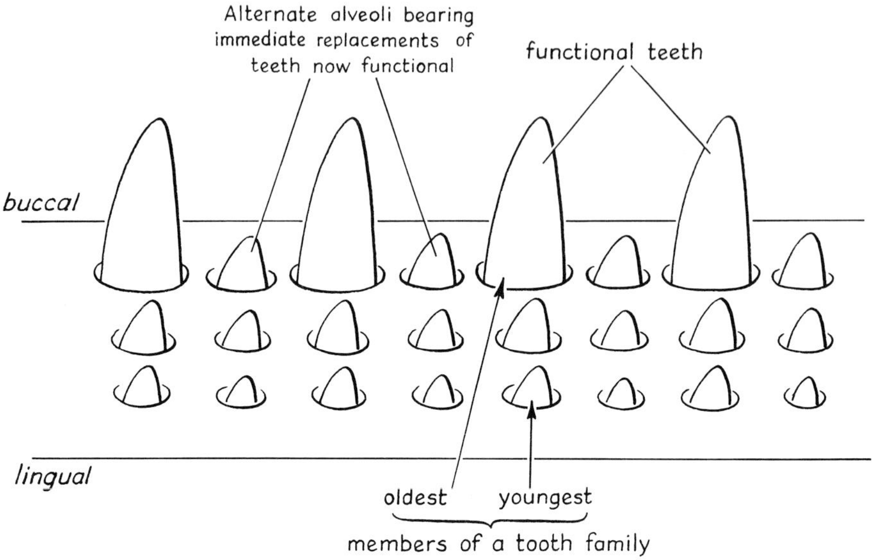


**Figure 5.2:** Diagrammatic Representation of Tooth Replacement in a Primitive Reptile
Source: After Kermack, 1956

## Evolution of Mammalian Canines

In the captorhinomorph cotylosaurs, which were carnivorous or insectivorous, the teeth are simple pointed cylinders, slightly recurved. Their function was simply to seize the prey, which was then either swallowed alive or banged repeatedly on the ground to quieten it and eventually kill it. Some of the present day monitor lizards (*Varanus*) show this behaviour. It is not efficient, and it is better that the prey, once seized, is killed quickly.

The ophiacodont pelycosaur *Ophiacodon*, which was possibly a fish-eater (Romer and Price, 1940), had a long tooth-row in both upper and lower jaws (Figure 5.3). The teeth themselves are long blades, they are slightly recurved and lack any accessory cusps. In the lower jaw there is a gradual increase in size of the teeth as we pass mesially, that is towards the symphysis, along the jaw. In the upper jaw there is a similar gradation in size except for a group of three or four teeth about a third of the way back from the front of the maxilla, which are the largest of all. In another ophiacodont, *Varanosaurus*, a pair of teeth occupying a similar position in the maxilla are the largest teeth in the whole jaw (Figure 5.3); and in *Sphenacodon* the whole of the maxillary tooth row in front of these teeth has so shortened that they lie close to the suture with the premaxilla (Figure 5.3). Finally, in *Dimetrodon*, the tooth row is further shortened and only one tooth is implanted in the maxilla mesially to this pair of large teeth (Watson, 1951). These teeth have now become so much larger than their fellows as to merit the designation *canines*.

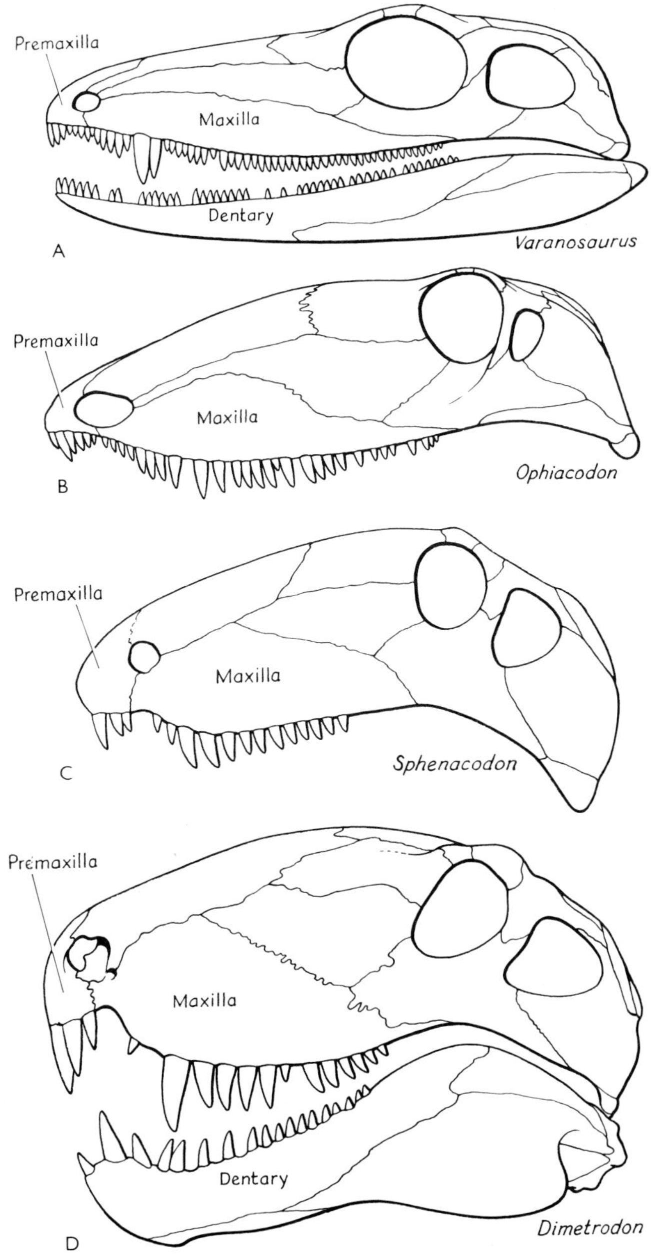


**Figure 5.3:** Skulls of Pelycosaurs to Show the Enlargement of the Teeth in the Anterior Part of the Maxilla to Form Incipient Canines
Source: After Watson, 1951

The teeth mesial to them can be termed *incisors* and those distal to them *cheek-teeth* (Figure 5.3). It is not possible to sub-divide these cheek-teeth into premolars and molars as this terminology implies a mode of tooth replacement characteristic of the Mammalia (Figure 5.9), which did not occur in these mammal-like reptiles.

Thus these pelycosaurs illustrate the early stages in the evolution of the characteristic mammalian pattern of dentition.

The canines of a carnivorous mammal, e.g. modern fissipedes such as the cats, are extraordinarily efficient for both seizing and killing prey. The deep penetration of the teeth into the body of the victim ensures a firm secure hold whilst the deep stab wounds are quickly fatal. Stabbing is a remarkably effective way of killing, as man himself has found out early in his history, perfecting such weapons as the dagger, spear and rapier. Much of the efficiency of these weapons depends upon their being used on the right place in the victim's body; carnivores show considerable skill in choosing the right place, cats and mustelids show a preference for biting at the nape of the neck, where damage to the vital base of the brain causes instant death. *Dimetrodon* may not have been such a sophisticated killer, but it certainly did not bite at random. It is worth mentioning that most carnivorous birds (diurnal birds of prey and the owls) also seize and kill by stabbing, although here the stabbing is carried out by the talons.

*Dimetrodon* had a pair of upper canines in each maxilla because the teeth were probably replaced fairly frequently; the animal could not risk being left with no upper canines at all on one side while replacement was taking place. These teeth were vital for obtaining food, so they were replaced as an alternate pair. Usually fossil specimens have two functional upper canines in each maxilla, but sometimes animals are found which have only one tooth of the pair functional at the time of death; the other was being replaced. While *Dimetrodon*, like the other advanced sphenacodonts, quite clearly had upper canines, the lower canines were much less defined. There are two or three enlarged teeth towards the anterior end of the dentary occluding in front of the upper canines. They are much less distinctly marked off from their neighbours than are the upper canines from theirs (Figure 5.3).

The paired upper canines of the sphenacodonts were obviously a less effective 'dagger' than the single upper canines of the modern fissipede. After all, a knife is a more effficient stabbing instrument than is a fork. So it is no surprise that the successors of the pelycosaurs have, with only few exceptions, but one functional canine in each maxilla. This they have acquired by lengthening the time for the next tooth of the same family to become functional after its predecessor has been shed. So unlike *Dimetrodon*, there is normally a single functional canine in each maxilla, with only one alveolus having the functional canine at any time and the other of the pair containing either the root of the previous canine undergoing resorption or a developing replacement tooth.

As well as showing the early stages in the evolution of the varied and differentiated mammalian tooth pattern from a simple reptilian one, these pelycosaurs also show continual and alternate tooth replacement typical of early tetrapods. In the latter the immediate functional replacement tooth tended to be the adjacent teeth on either side and not another tooth in the same 'socket' or alveolus. (A diagrammatic representation of this is given in Figure 5.2).

Each alveolus bears a 'family of teeth' which succeed one another continuously, so whilst one series of teeth is functional, the other is in the process of replacement.

In Figure 5.2 each series is shown replacing teeth simultaneously, but in practice there is often a constant time difference between the replacement of adjacent alveoli in the same series. This 'phase difference' produces the effect of 'waves' of replacement passing backwards and forwards along the jaw. If any given alveolus shows replacement before the next member of the series distal to it, the 'wave' will pass distally. If the alveolus shows replacement after its distal neighbour the 'wave' will run mesially towards the jaw symphysis. For strict alternation the phase angle between the two series must be 180° and any other phase difference between adjacent members of the same series will tend to obscure the distichial nature of the replacement teeth, particularly in a short jaw (Edmund, 1960).

The teeth of *Dimetrodon* are structurally almost identical, being slightly recurved and rather cones with the differentiation based on size and function (Figure 5.3). *Dimetrodon* is a good example of a carnivore, which uses the largest teeth in the upper jaw for seizing, holding and killing prey. These teeth are the canines and the teeth mesially to them, the incisors, are used for biting and tearing off lumps of flesh. The post-canines hold the food whilst it is being swallowed but do not cut it prior to swallowing. This arrangement of the teeth is exactly paralleled in the modern cat, where like *Dimetrodon*, the incisors are borne on the premaxilla and the canines and post-canines are inserted in the maxilla. In the modern cat, however, the post-canines are the teeth which cut up the prey.

The lower jaw of *Dimetrodon* occasionally, but not always, shows comparable development of the lower canines. In *Dimetrodon limbatus* (specimen No. MCZ 1347, Romer and Price, 1940) the second tooth from the mesial end of the tooth row is much the largest in the lower jaw and can merit the title of 'canine'. There is a pair of upper canines in each maxilla but only one lower canine in each dentary. The lower canine does not seem to have been a member of a distichial pair. In most sphenacodonts, the lower canine is not as clearly differentiated from the rest of the teeth in the dentary as it is in the specimen of *Dimetrodon limbatus* cited. Generally the lower jaw seems to lag behind the upper in the differentiation of the teeth.

Moving from the pelycosaurs to the therapsids, the dentition and method of tooth replacement have been studied extensively in the Gorgonopsia and

Therocephalia; the relative ease with which their skulls can be developed in acid from the matrix in which they were fossilised has assisted this study.

In some Gorgonopsia, the upper canine is immense and for such a tooth to develop there must be as much or even more enlargement of the root below the gum-line as there is in the crown above it. This produces consequential changes in the shape of the maxilla and also in the teeth in the rest of the tooth row. *Rubidgea atrox* from the *Cistecephalus* zone of South Africa is an example of a gorgonopsid with giant upper canines, but the upper jaw has only one tiny cheek-tooth and there are only two in the lower jaw.

Similarly elongated upper canines are found in a number of different groups of carnivorous mammals, where they probably evolved quite independently. The best-known are the large sabre-toothed cats (Machairodontinae) such as *Smilodon* (Figure 5.4) and similar large canines are present in the Eocene creodonts, *Apataelurus, Machairoides and Anacodon*. The first two are related, but *Anacodon* acquired its long upper canines quite independently and none of the creodonts are in any way related to the sabre-toothed cats. Similar long upper canines evolved in the Pliocene South American marsupial *Thylacosmilus*.

Simpson (1941) showed that the function of the long upper canines was to enable the carnivore to stab and kill large prey with a hard thick skin or integument. It would have been difficult to attack such a prey satisfactorily by simply biting it. The present-day walrus has similar elongate upper canines,

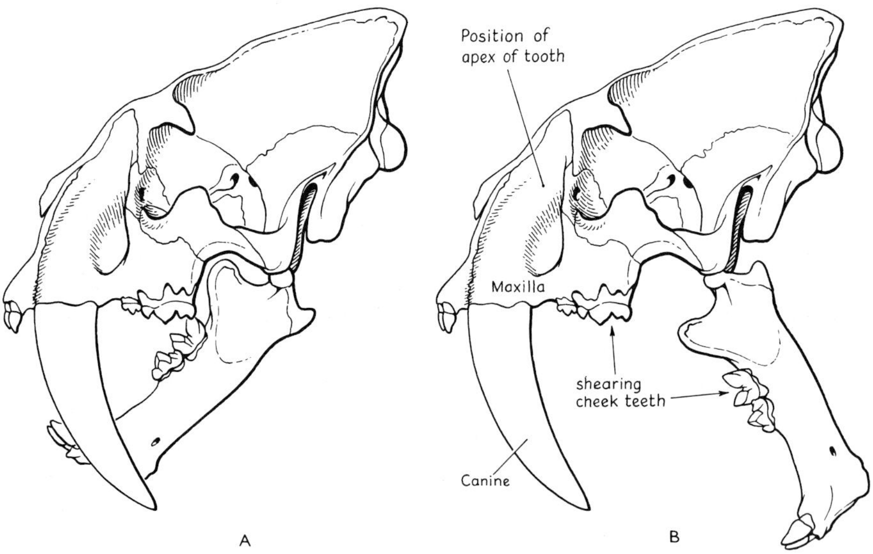


**Figure 5.4:** The Skull of *Smilodon* Showing the Extreme Gape of the Lower Jaw to Clear the Tips of the Enlarged Upper Canines
Source: After Simpson, 1941

with which it is said to kill seals by downward stabbing (Pycraft, 1928). In all these mammals the back part of the lower jaw becomes modified to permit extreme opening of the mouth to clear the tips of the upper canines and for the same reason the lower canines are greatly reduced in length and play no part in the capture of the prey. The stabbing movement is made by the whole head, which is pulled down by the powerful neck muscles attached to the back of the skull. The jaw-closing muscles are weak. In all except *Anacodon* and the walrus, the cheek-teeth show carnassial modification, so they can act as a pair of shears cutting up the flesh and bone of the stabbed prey.

*Rubidgea* does not fit into this pattern. This gorgonopsid had long lower canines as well as upper canines as elongate as any of the mammals mentioned above. How *Rubidgea* dealt with its food and how it used its long canines is a mystery. It would have been difficult for it to have stabbed its prey with the upper canines as the mouth would have had to have been opened unbelievably wide to clear the tips of long lower canines. The reduction in its cheek-teeth indicate that it must have relied upon its incisors to tear chunks off its prey and then swallowed them whole as in all gorgonopsids and therocephalians.

In some gorgonopsids with long upper canines there is a small flange on the lower jaw to protect the tip of the upper canine when the jaw is closed. This is shown well in *Inostrancevia*, the largest known gorgonopsid from Russia. There is a similar flange in the mammalian carnivores with long upper canines, but *Rubidgea* has no such protection for its canines.

The upper canines of *Aelurosaurus felinus* show functional alternate replacement (Figure 5.5.). On either side of the skull there were two alveoli or sockets, one behind the other, and each could bear a functional canine tooth. This figure shows that in this particular specimen the functional canine is borne in the more anterior of the pair of alveoli on each side of the skull. On the right side, the immediate replacement of the functional canine is a large tooth filling the posterior alveolus. In its turn, this tooth would have then be replaced by a tooth of the tooth-family of the anterior alveolus and this can be seen as a small tooth lying lingually to the now functional canine. On the left-hand side, the posterior alveolus is filled with the remains of an old root, being invaded by spongy bone and in the process of resorption. This indicates that the upper canines were shed by the tooth being eroded along the gum-line with the crown subsequently breaking off and the root resorbing. Lingually to the old resorbing root lies a replacement tooth, which is a member of the same tooth-family and, had the animal lived, would have been the immediate successor to the functional canine. It is much smaller than its fellow on the opposite side of the mouth, indicating the upper canines were not shed at the same time. The presence of the remains of the old root of the immediate predecessor of the functional canine confirms that the right functional canine is the older.

It is not known why or how this method of shedding and replacing canine

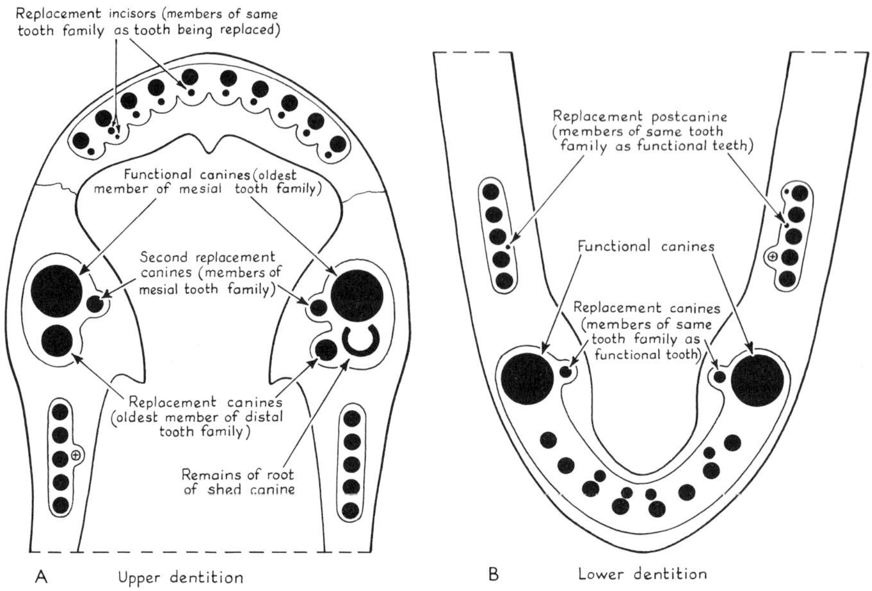


**Figure 5.5:** Tooth Replacement in *Aelurosaurus felinus*
Source: After Kermack, 1956

teeth evolved. It may be associated with the hazard of infection of a deep alveolus had the canine tooth been shed in the normal way. The resorption of at least the root of a tooth reduces the mineral loss from the animal's body that otherwise would have occurred had the large tooth just dropped out of its socket when it became too worn to be useful.

Figure 5.5 shows that in the upper jaw of *Aelurosaurus felinus* all the incisors were being replaced, as there is a small replacement tooth lying lingually to the tooth which it has to replace. Two replacement upper cheek-teeth can be seen on each side of the skull. In the lower jaw there are four replacement incisors, one replacement cheek-tooth on the left and three on the right. Each lower canine bears a replacement lingually to it. Thus the replacement of all these teeth is quite different from the functionally distichial pattern of tooth replacement found in the Pelycosauria, the primitive Amphibia and in the primitive reptiles. In *Aelurosaurus* and the other gorgonopsids alternate replacement is only retained in the upper canines. In the other teeth any alternate replacement is quite obscured by the shortness of the tooth-row, and the replacement tooth is a member of the same tooth-family as the functional tooth it replaces. The retention of alternate replacement in the upper canines is undoubtedly due to their great size and their importance to the animal. With such large teeth replacement must have been a slow business, and the animal might well not have been able to have made a satisfactory kill of its prey with the upper canine missing on one side of the jaw.

In many specimens of gorgonopsids, there is no sign of tooth replacement or only a limited amount, often on one side only. In such cases the upper canine on each side is in the anterior alveolus, with a plug of spongy bone visible behind it. Although there appears to be no visible tooth replacement, dissection of the plug of bony tissue filling the alveolus, sectioning the skull or radiography of a skull prepared in acid will sometimes reveal the remains of an old root within the plug. The figure of *Scylacops capensis* (Figure 5.6) illustrates this: here tooth replacement is occurring only in the lower incisors and cheek-teeth, there being no replacement at all taking place in any of the upper teeth. Both functional canines are borne in the anterior of each pair of alveoli, whilst the posterior alveoli are filled with a plug of spongy bone, within which are the roots of the penultimate canines. The root on the left is the older with the spongy bone invading its anterior wall as far as the pulp cavity. The canine on the right was shed much nearer the death of the animal, as this root is much less eroded and its pulp cavity is only partially filled with spongy bone. Lying buccally and anteriorly to this root is a piece of yet another tooth, which can only be the remains of the antepenultimate canine on this side of the jaw.

Table 5.1 gives details of the dentition of ten gorgonopsid and six therocephalian skulls. Table 5.2 gives data for the upper canines which assumes, as seems perfectly reasonable, that replacement occurs independently on the two sides of the jaw. Thus the 16 specimens produced between them 27

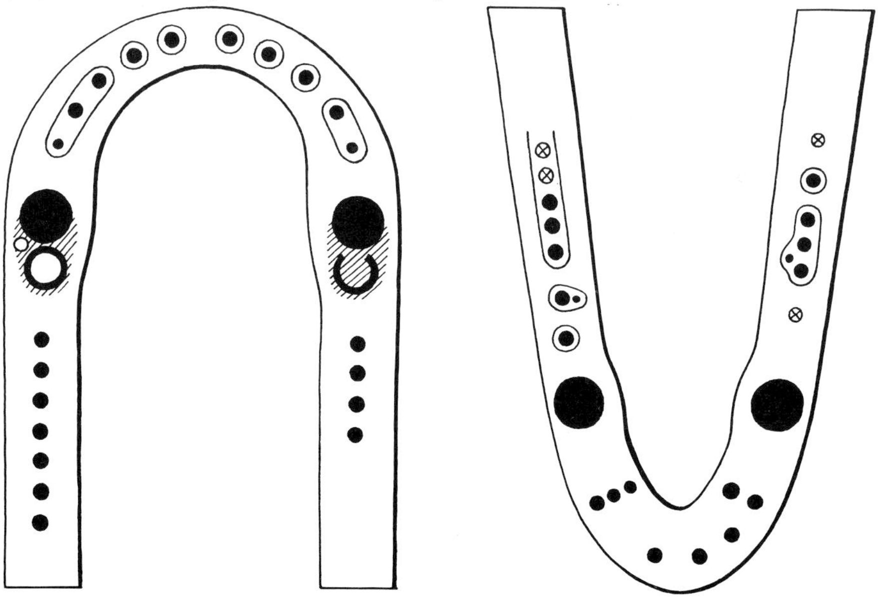


**Figure 5.6:** Tooth Replacement in *Scylacops capensis*
Source: After Kermack, 1956

| | | GORGONOPSIA | | | | | | | | | | THEROCEPHALIA | | | | | |
|---|---|---|---|---|---|---|---|---|---|---|---|---|---|---|---|---|---|
| | | *Aelurosaurus felinus* (R 339) | *Aelurosaurus felinus* (R 855*a*) | *Aelurosauroides watsoni* (R 855) | an aelurosaurid (337) | *Dixeya quadrata* (79) | *Gorgonops torvus* (R 1647) | *Cynariops robustus* (R 5743) | *Scylacops capensis* (R 121), partly sectioned | *Scymnognathus whaitsi* (R 49396), sectioned specimen | large unidentified gorgonopsid (R 256), sectioned specimen. Right side only preserved | *Trochosaurus major* (R 5747) | *Pristerognathus polyodon* (R 2581) | *Pristerognathus polyodon* (R 393) | *Cynariognathus platyrhinus* (R 4097), sectioned specimen | *Notosollasia laticeps* (R 5699) | *Whaitsia* sp. (100) |
| UPPER DENTITION | | | | | | | | | | | | | | | | | |
| dental formula ... | | 5.1.5 | 5.1.5 | 5.1.5 | 5.1.5 | 5.1.4 and 5 | 5.1.? | 4.1.3 | 5.1.7 | ?.1.5 or 6 | ?.1.? | 5.2.3? | 6 and 7. 1.4+ | 6.1.3+ | 6?.1.6 | 5.1.0 | 5.1.0 |
| teeth in process of replacement: incisors | R | all | 1, 3, 5 | none | all | none | none | 2, 3, 4 | none | — | — | none | none | 1 | —* | none | none |
| | L | all | 2, 3, 5 | none | all | none | none | 2, 3 | none | — | — | 5 | none | 2 | none | none | none |
| canines | R | yes | yes | no | yes | no | no | yes | no | no | no | both | no | no | —* | no | no |
| | L | yes | yes | no | yes | no | no | obs. | no | no | — | obs. | no | no | no | no | no |
| cheek teeth | R | none | 4 | none | 1 | none | none | none | n.v. | n.v. | — | obs. | none | none | —* | — | — |
| | L | none | 4 | none | 1, 2 | none | none | none | n.v. | n.v. | — | ? | 3 | none | none | — | — |
| position of plug of spongy bone in the pair of canine alveoli on each side | R | abs. | abs. | post. | ant. | post. | post. | obs. | post. | post. | post. | abs. | post. | post. | —* | post. | post. |
| | L | abs. | abs. | abs. | ant. | post. | post. | abs. | post. | post. | — | obs. | post. | post. | post. | post. | post. |
| position of root of old canine in pair of canine alveoli | R | abs. | ant. | abs. | ant. | post. | post. | obs. | post. | post. | post. | abs. | post. | post. | —* | obs. | post. |
| | L | post. | abs. | post. | ant. | post. | post. | post. | post.† | post. | — | obs. | post. | post. | post. | post. | post. |
| LOWER DENTITION | | | | | | | | | | | | | | | | | |
| dental formula ... | | 4.1.5 | 4.1.5 | — | 4.1.5 | — | — | — | 4.1.7 | 4.1.4 | 5.1.2+ | 3.1.5 | 3.1.7 and 8 | 3.1.7 | 1+.1.8 | 0.1.0 | — |
| teeth in process of replacement: incisors | R | 1, 4 | 1, 2 | — | 1, 2, 3 | — | — | — | none | 3 (poss.) | none | 1 | none | none | n.v. | — | — |
| | L | 1, 2, 3 | 1, 3 | — | 1, 2, 3 | — | — | — | none | n.v. | — | 1, 3 | 2 | none | n.v. | — | — |
| canines | R | yes | yes | — | yes | — | — | — | none | none | none | yes | none | none | n.v. | none | — |
| | L | yes | yes | — | obs. | — | — | — | none | none | — | yes | none | none | n.v. | none | — |
| cheek teeth | R | 3 | 2 | — | none | — | — | — | 2 | none | — | not known | none | none | n.v. | — | — |
| | L | 3 | 3, 5 | — | 3 | — | — | — | 3 | none | — | not known | 6 | none | n.v. | — | — |
| roots of old canines | | none | none | — | none | — | — | — | none | none | none | none | none | none | n.v. | one on each side | — |

*Key.* abs. = absent, post. = posterior, ant. = anterior, obs. = obscured, n.v. = none visible, poss. = possibly.

* Upper jaw dentition preserved on left side only.

† Small piece of root of another canine lies buccal to right functional tooth.

*Note.* This table is a record of replacement teeth actually present in the specimens. No mention has been made here of empty alveoli from which replacement teeth may have fallen after the death of the animal.

**Table 5.1:** Tooth Replacement in Gorgonopsids and Therocephalians

alveolar pairs, as in some of the specimens one side of the jaw was either missing or was so damaged that the condition of the alveoli could not be assessed. In the seven pairs where replacement was still occurring, the functional canine was in the anterior position in three and in the posterior one in four. This is just what would be expected on the hypothesis of the continuous alternate replacement of the canine teeth.

The twenty other pairs of alveoli, in which no replacing canines are present, show a very different picture. Here the functional canine is always to be found in the anterior alveolus. This cannot simply be due to chance, since the $\chi$ squared test shows that the probability of this is 0.002. It can be explained, however, if it is postulated that when the animal reached a certain age replacement of the upper canines ceased. Since the final functional upper canines is always found in the anterior alveolus, this also strongly suggests that the replacement was not only limited, but also occurred a definite number of times (Kermack, 1956).

During the growth of the animal the upper canines would have to be replaced a number of times to keep the size of the tooth matched to the size of the skull — a point that is illustrated by the diameter of the root of the functional canine being larger than that of its predecessor in almost all cases. Thus when the growth of the animal ceased, replacement of the canines also ceased. There were advantages in this. Having dispensed with the unpleasant business of losing an important part of its dental battery at intervals, the animal retained an appropriately sized canine, which was now held firmly in its socket. This last point can be illustrated by comparing specimens showing continuous tooth replacement with those in which this had ceased. In the former the functional canine fitted so loosely in its alveolus that it is often missing in the fossil, in the latter the bone of the socket grasps the tooth firmly, and even if the crown has been broken off the root is retained.

It is tempting to compare the limited, definite replacement in the Gorgonopsia and in the Therocephalia with the limited definite replacement of the teeth of placental mammals. In the latter all the teeth are replaced once with the exception of the molars, which are, of course, never replaced. The first set of teeth is known as the milk, or deciduous dentition, the second set being the permanent teeth. Bolk (1922), on other grounds, actually postulated that the deciduous and permanent dentitions of mammals were the survivors of the two alternately replaced series found in reptiles. This seems improbable.

In the specimens in which replacement of the upper canines had ceased, replacement in the other teeth was either greatly reduced or was absent (Table 5.1). It is a reasonable assumption that replacement eventually ceased in these teeth, as we know it did in the upper canines. It is not known if this replacement was also definite, in the sense that each alveolus bore a determinate number of teeth, but it is likely. Whereas it is not unusual for tooth replacement in modern reptiles to peter out and finally stop, as the animal ages — *Sphenodon* is a good example of this — but this is not the same as in

| | | | Gorgonopsia | | | | | | | | | | Therocephalia | | | | | |
|---|---|---|---|---|---|---|---|---|---|---|---|---|---|---|---|---|---|---|
| | | | *Aelurosaurus felinus* (R 339) | *Aelurosaurus felinus* (R 855 *a*) | *Aelurosauroides watsoni* (R 855) | an aelurosaurid (337) | *Dixeya quadrata* (79) | *Gorgonops torvus* (R 1647) | *Cynariops robustus* (R 5743) | *Scylacops capensis* (R 121), partly sectioned | *Scymnognathus whaitsi* (R 49396), sectioned specimen | large unidentified gorgonopsid (R 256) | *Trochosaurus major* (R 5747) | *Pristerognathus polyodon* (R 2581) | *Pristerognathus polyodon* (R 393) | *Cynariognathus platyrhinus* (R 4097) sectioned specimen | *Notosollasia laticeps* (R 5699) | *Whaitsia* sp. (100) |
| anterior canine alveolus | functional teeth | R. | + | 0 | + | 0 | + | + | 0 | + | + | + | + | + | + | – | + | + |
| | | L. | + | + | + | 0 | + | + | + | + | + | – | + | + | + | + | + | + |
| | replacing teeth | R. | + | + | 0 | + | 0 | 0 | + | 0 | 0 | 0 | + | 0 | 0 | – | 0 | 0 |
| | | L. | + | + | 0 | + | 0 | 0 | ? | 0 | 0 | – | ? | 0 | 0 | 0 | 0 | 0 |
| | plug of spongy bone | R. | 0 | 0 | 0 | + | 0 | 0 | ? | 0 | 0 | 0 | 0 | 0 | 0 | – | 0 | 0 |
| | | L. | 0 | 0 | 0 | + | 0 | 0 | 0 | 0 | 0 | – | ? | 0 | 0 | ? | 0 | 0 |
| | old root | R. | 0 | + | 0 | + | 0 | 0 | ? | + | 0 | 0 | 0 | 0 | 0 | – | 0 | 0 |
| | | L. | 0 | 0 | 0 | + | 0 | 0 | 0 | 0 | 0 | – | ? | 0 | 0 | 0 | 0 | 0 |
| posterior canine alveolus | functional teeth | R. | 0 | + | 0 | + | 0 | 0 | + | 0 | 0 | 0 | + | 0 | 0 | – | 0 | 0 |
| | | L. | 0 | 0 | 0 | + | 0 | 0 | 0 | 0 | 0 | – | + | 0 | 0 | 0 | 0 | 0 |
| | replacing teeth | R. | + | 0 | 0 | 0 | 0 | 0 | ? | 0 | 0 | 0 | + | 0 | 0 | – | 0 | 0 |
| | | L. | + | + | 0 | 0 | 0 | 0 | ? | 0 | 0 | – | ? | 0 | 0 | 0 | 0 | 0 |
| | plug of spongy bone | R. | 0 | 0 | + | 0 | + | + | ? | + | + | + | 0 | + | + | – | + | + |
| | | L. | 0 | 0 | 0 | 0 | + | + | 0 | + | + | – | ? | + | + | + | + | + |
| | old root | R. | 0 | 0 | 0 | 0 | + | + | ? | + | + | + | 0 | + | + | – | ? | + |
| | | L. | + | 0 | + | 0 | + | + | + | + | + | – | ? | + | + | + | + | + |

**Table 5.2:** Canine Replacement in Gorgonopsids and Therocephalians

these mammal-like reptiles, where the determinate limited tooth replacement characteristic of modern mammals first appears (see Kermack, 1956 for a full discussion).

## The Evolution of Mammalian Cheek-Teeth

The prey once dead, the next problem was to eat it. A modern fissipede carnivore, such as a cat or a dog, cuts up its prey with its powerful cheek-teeth. These carnassial teeth act as shears, cutting flesh and bone alike. Gorgonopsids and Therocephalia had nothing equivalent to these. Their cheek-teeth were simple, small, peg-like teeth and quite incapable of cutting anything. Once in the mouth the lump of flesh was swallowed whole. The tearing of the lump of flesh from the body of the prey was done by the incisors. These were long, powerful teeth, quite unlike the tiny incisors of a modern fissipede, where they are used mainly for the toilet, as may be seen by watching a cat washing itself. After its prey was dead, one of these therapsids must have held the body down firmly with the front paws, seized a lump of flesh with the incisors, and then by powerful side by side movements of the head torn the lump from off the carcass. Once this lump was in the mouth it was of necessity swallowed whole, since there was no possibility of dividing it further. The small cheek-teeth would merely have served to guide the lump of flesh in its passage backwards in the mouth.

This method of eating, although effective, was slow. It shows analogies with that employed by birds of prey at the present day. Whilst these kill their victims as quickly as a fissipede by stabbing with the talons, the subsequent eating of the prey is a slow and laborious process. The bird holds the carcass firmly with one foot, while tearing off lumps of flesh with its bill, using side to side movements of the head, just as did the therapsids already mentioned. Not only is this slow, but there is no way in which the bird can break up the bones of the prey and reduce them to a size small enough to be swallowed. This part of the carcass is wasted. It is instructive to watch two animals of comparable size, a kestrel and a ferret, eat a mouse. The bird patiently picks the flesh off the bones, while the mammal rapidly devours the whole mouse, bones, fur and all, crunching it up rather like a man eating an apple. The difference is striking: there would be a similar striking difference if one were able to watch a gorgonopsid or a therocephalian eat its prey and compare it with a fissipede.

Turning now to the cynodonts, these were the dominant therapsid carnivores of the Lower Trias. They possessed similar, but admittedly less spectacular canines to those found in the gorgonopsids, but they functioned in the same general way in killing the prey. The cheek-teeth are different from those in the gorgonopsids and in the Therocephalia in showing, for the first time, the presence of secondary cusps. The cynodonts retained the large canines of the Permian therapsids and must have torn their prey to pieces in the same way. They probably could not, in most cases, have cut up the food in the mouth with the cheek-teeth. The use of the cheek-teeth to cut up food poses certain mechanical problems which must be considered.

Two terms which we shall use in describing the action of the teeth need definition. One, 'to shear' or 'shearing', is no problem. Teeth shear when the opposing teeth cut like the blades of a pair of scissors; the carnassial teeth of a cat form an excellent example. The other term is 'grind' or 'grinding' and is more difficult. In common usage 'grind' has two quite different meanings. We say that we grind a substance in a pestle and mortar. Here we mean that the substance is being smashed and bruised to a fine powder or to a pulp, depending upon its brittleness. The other meaning is exemplified when we talk about grinding a tool on a grinding wheel. In this case we are really talking about microshearing by the tiny sharp points of the crystals of aluminium oxide or silicon carbide in the wheel. In the USA meat is ground in a mincer — again shearing. In respect to teeth it is almost always the second meaning that is meant. Actually 'rasping' would be a better term to use, as grinding teeth function exactly as two rasps opposed to each other. The term 'grinding' is hallowed by long usage and cannot be avoided, but care must be taken to remember exactly what it means.

In cutting up the food the cheek-teeth function as a pair of shears or as the blades of a pair of scissors. When a pair of scissors is closed in cutting, the material being cut tends to go between the blades and twist them around their long axes. This twisting would separate the cutting edges and render their action ineffective were it not prevented by the action of the pivot (Figure 5.7). It is common knowledge that if the pivot is loose, then the cutting edges can rotate away from each other and the scissors will not cut. It is also common knowledge that it is quite remarkably difficult to cut the nails of the right hand with a pair of scissors held in the left. The reason is that the scissor blades need to be held firmly in apposition against the wedging action of the material between them. Although both blades of a pair of scissors are identical, each individual blade has no plane of symmetry. Thus it is possible to have a pair of similar blades which are their mirror images. A normal, or right-handed, pair has two right-handed blades and the thumb is normally placed uppermost with the fingers below. The thumb pushes and the fingers pull, so that when the scissors are held in the right hand this action presses the blades together; but when the scissors are in the left hand, the blades are pulled apart and cut less efficiently. With scissors having a pair of left-handed blades the converse is true, as it is if the scissors are held with the fingers up and the thumb down.

A mammal or mammal-like reptile using its cheek-teeth to shear up its food must be able to solve the same two problems: firstly, it needs to hold the shearing teeth in occlusion against the wedging action of the food tending to separate them; and secondly the jaw joint must be able to resist without

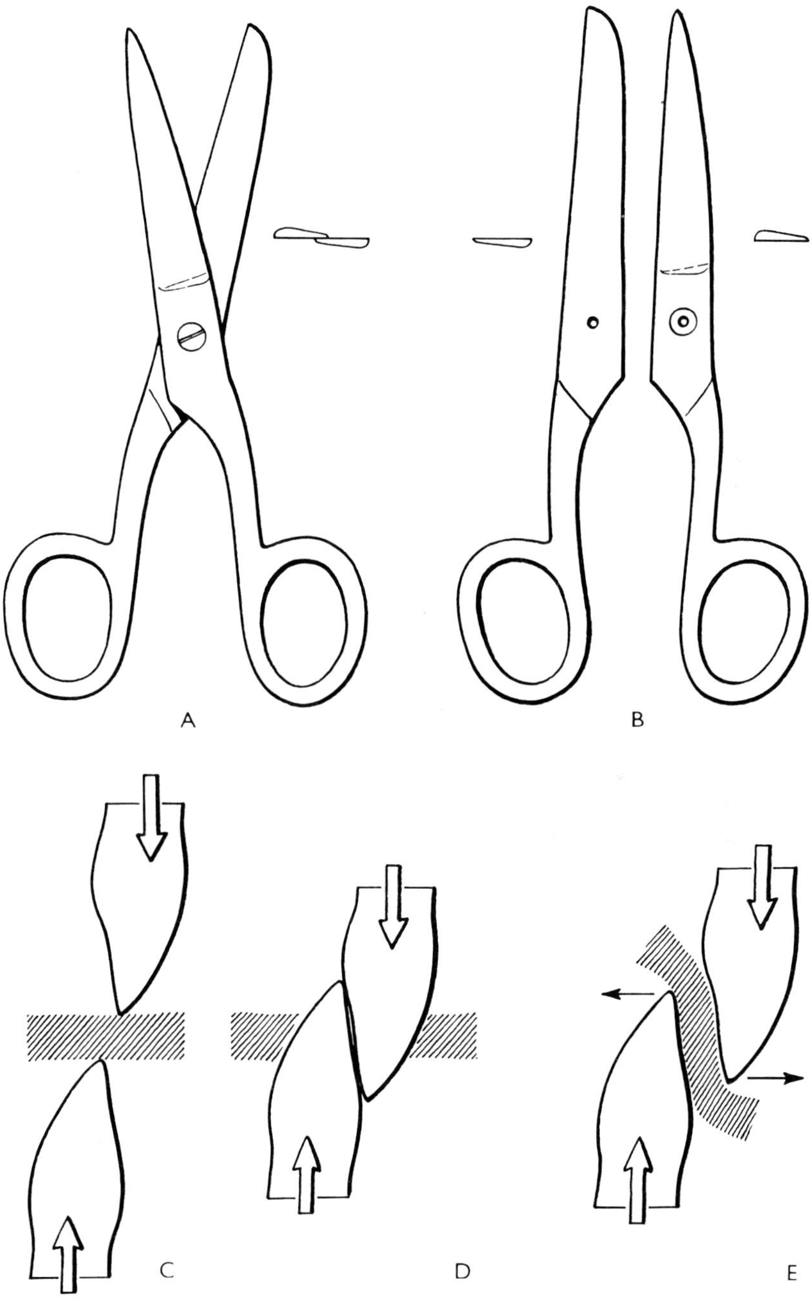


**Figure 5.7:** The Action of a Pair of Scissors. In D the pivot is firm and in E slack

dislocation the torques produced by the same wedging action. The torques will tend to produce rotation at the jaw joint which tend to dislocate it. The rotations are about the long axis of the jaw and about a vertical axis through the jaw articulation. The extent of the problem can be seen by examining the jaw articulation of a cat or badger, where all the rotations except those in the desired plane (opening and closing of the jaws) are completely suppressed. The analogy with the blades of a pair of scissors is plain.

The first problem is simply solved by the teeth being firmly held in occlusion by the muscles which close the jaws together. Obviously the post-canine teeth can only be in occlusion, that is be shearing, on one side at a time. The reader can confirm this every time he eats a meal, or more dramatically by eating an apple in front of a mirror. Obvious this may be, but it seems to have escaped the attention of most palaeontologists until 'active occlusion' was pointed out by Professor Bryan Patterson in 1956. Until then teeth had always been assumed to work in 'centric occlusion', that is with the mid-line of the lower jaw in the same vertical plane as that of the upper jaw, and with the lower teeth being symmetrically arranged with respect to those of the upper jaw.

The powerful jaw-closing muscles of the carnivorous cynodonts would have held the cheek-teeth in active occlusion without difficulty. It is the second problem of preventing undesirable rotations which largely defeated them. Cynodonts have a powerful and well-formed jaw articulation, quite as strong as is that in both the Gorgonopsia and Therocephalia. Theories which suggest a progressive reduction of the jaw articulation simply do not correspond with the facts (Crompton, 1963). The accessory jaw bones, namely the bones of the lower jaw other than the dentary, are indeed greatly reduced in size, but the jaw articulation itself is not (Kermack *et al.*, 1973). The reduction in size of the accessory jaw bones will be discussed under hearing (Chapter 6). The problem for the cynodonts was simple: the jaw hinge, powerful though it was, lay on both the axes around which rotation had to be prevented; the longitudinal axis of the jaw running along the base of the lower teeth, and the vertical axis passing through the jaw articulation itself. But to resist these rotations a thrust bearing is necessary remote from both axes. Crompton (1972) has shown that in some of the cynodonts there was a rudimentary bearing of that type. It lay lateral to the main jaw articulation, as was necessary for it to function, and it was formed by the contact of the surangular and the squamosal. This restraint was not particularly efficient, and the shearing action of the cheek-teeth must have been of only limited efficiency. In one group of cynodonts studied by Romer (1969, 1970), the Probainognathidae, a powerful thrust bearing had formed between squamosal and the surangular, but there may possibly have been a small involvement of the dentary itself as well. It is significant that in *Probainognathus* and its allies the cheek-teeth show heavy wear comparable with that of a mammal. Despite such tentative experimental moves in the right direction, the real solution of this problem lay with the ancestors of the mammals. In truth, it was their solution of the problem which turned them into mammals.

The dentition of the cynodont *Diademodon* indicates that it was able to both cut up and grind food in the mouth with is cheek-teeth. The upper dentition is shown in Figure 5.8 (Hopson, 1971). Behind the large canines is a gap or diastema and then a row of 13 cheek-teeth. These consist of four simple conical teeth, a further six more complicated teeth, which increase in size distally along the jaw, then a transitional eleventh post-canine followed by two elongated teeth with blade-like sectorial crowns. The six complicated teeth were grinding teeth, with flattened, transversely widened crowns. In those mammals (bears and badgers are good examples) where there are both shearing and grinding teeth, the sectorial teeth precede the crushing teeth in the jaw. *Diademodon* has its teeth arranged exactly the other way round. Bears and badgers are truly omnivorous, eating both animal and vegetable prey. They use their shearing carnassial teeth to cut up flesh, and their flat crowned molars to mince vegetable matter. Since both processes cannot occur at the same time, it is probably quite immaterial whether the grinding teeth come before the sectorial teeth or vice versa. Thus *Diademodon* had a perfectly functional dentition to deal with an omnivorous diet.

If this complex dentition were necessary for the nutrition of the animal, then, unless there were a definite pattern of tooth replacement, there would be dificulties during growth. Fourie (1963) found that the smaller, and so presumably younger individuals of *Diademodon* had approximately the same number of cheek-teeth as the larger adult animals. In all the specimens there is the same differentiation into anterior cones, intermediate broad-crowned teeth, and posterior sectorial teeth. To keep the diastema the same relative size during growth, teeth were lost at the mesial end of the tooth row, while further sectorial teeth were added at its distal end. To maintain a

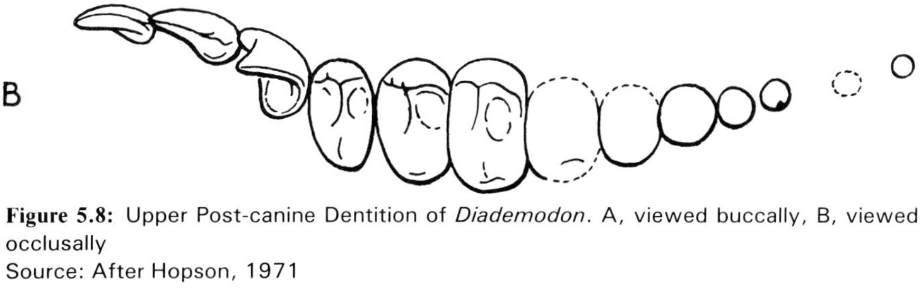


**Figure 5.8:** Upper Post-canine Dentition of *Diademodon*. A, viewed buccally, B, viewed occlusally
Source: After Hopson, 1971

balanced dentition, the anterior broad-crowned teeth were replaced by cones, whilst the mesial sectorial teeth were replaced by broad-crowned teeth. This replacement of a tooth by a tooth characteristic of a more mesial position in the jaw is typical of placental mammals (Figure 5.9), where it occurs for exactly the same reason as in *Diademodon*. Equally typical is the completion of the final dentition by the addition, at the distal end of the tooth row, of teeth which are never replaced.

It is highly improbable that *Diademodon* was in any but a most distant way connected with the ancestry of the mammals, or that the condition of tooth replacement in *Diademodon* and its allies gave rise directly to that of the placental mammals. What we are seeing here is the evolution of similar mechanisms to cope with similar problems. The common problem here is the

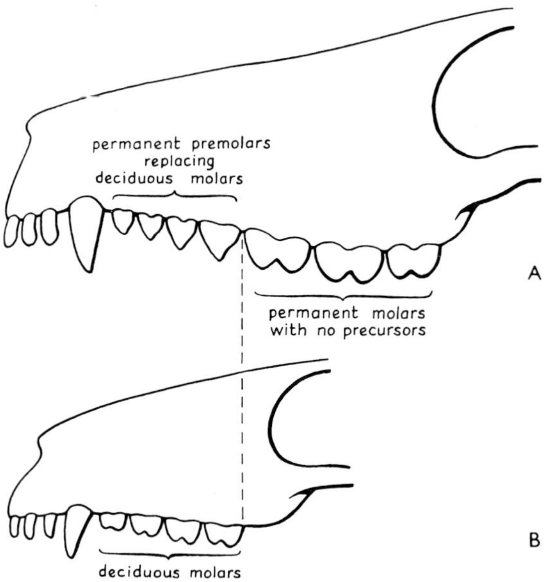


**Figure 5.9:** A, Permanent Dentition of a Placental Mammal; B, Milk Dentition

maintenance of a differentiated dentition through infancy and adolescence to the adult state.

Crompton (1963) has shown a similar pattern of tooth replacement in the Lower Triassic cynodont *Thrinaxodon*. Here the superficial appearance is even more placental-like, as the anterior cheek-teeth were replaced with by others with less complex crowns. This is reminiscent of the replacement of the more complex milk-molars of the placentals by the simpler premolars (Figure 5.9). This similarity in patterns of tooth replacement in *Thrinaxodon* and the placental mammals is as much a case of convergence as is the condition in *Diademodon*. In all cases there is a need to maintain a differentiated dentition during growth, and it is this which has led to the similar patterns of tooth replacement.

## The Bones of the Lower Jaw

Before going on to discuss the mammals themselves, the reduction in the size of the accessory jaw bones in cynodonts and in early mammals will be considered. In the earlier mammal-like reptiles these bones are large and form sutural contacts with the dentary and with each other. In the cynodonts and in the first mammals, however, the accessory jaw bones are greatly reduced in size and there is no sutural connection with the dentary. Instead they lie in a groove in the latter bone, and are bound in place by connective tissue (Figures 5.10 and 5.11). Although the accessory bones are, in general, greatly reduced in size, the jaw articulation shows only slight reduction. These changes are mainly, but not entirely, adaptively related to hearing (Kermack *et al.*, 1973, 1981; Allin, 1975), and will be referred to again in Chapter 6.

The other reason for the changes in the accessory jaw bones and their mode of attachment to the dentary is connected with the changes taking place in the cheek-teeth. In the earlier therapsids the teeth simply had the functions of seizing and killing the prey, and then tearing lumps of flesh off from the

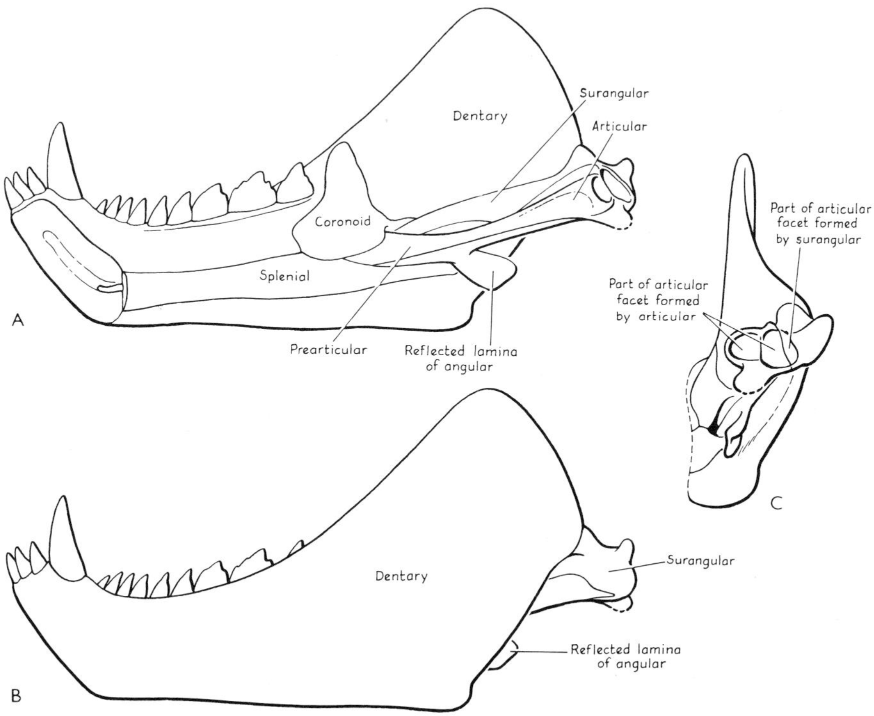


**Figure 5.10:** Lower Jaw of *Cynognathus*. A, medial view; B, lateral view and C, posterior view. See Figure 4.5 to 4.8 for the skull and compare with Figure 4.9 — the lower jaw of *Morganucodon*

carcass, the cheek-teeth doing no mastication or cutting. In the cynodonts the post-canines had acquired the additional function of cutting up, to a limited extent, the food in the mouth. Fissipedes, which also do this, have the problem that the canines prevent the proper or active occlusion of the cheek-teeth, when they themselves come into centric occlusion. This the canines have to do to achieve the maximum stabbing effect. To enable the canines to clear each other so that the cheek-teeth on one side come into active occlusion, the jaw makes a slight rotation about its longitudinal axis. This rotation is made possible by the median symphysis between the two rami of the lower jaw remaining unossified. Until the necessity of this rotation was realised, the failure of this symphysis to ossify in these carnivores, which surely needed a strong lower jaw, was a complete mystery. Some other

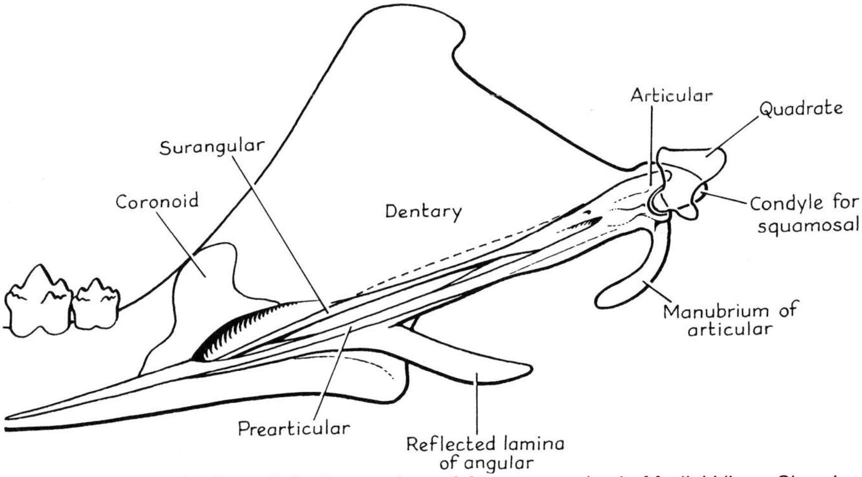


**Figure 5.11:** Posterior Part of the Lower Jaw of *Morganucodon* in Medial View, Showing the Double Jaw Articulation and Accessory Bones Lying in the Groove in the Dentary — see also Figure 4.9

mammals, particularly some insectivores, have the same problem which they solve in the same way. In the Cynodontia the two halves of the lower jaw also remain unossified at the symphysis, so that each dentary could rotate slightly on the accessory bones lying in the groove on its median side. This movement would have enabled the canines to clear each other and allowed the cheek-teeth to function to a limited extent as shears to cut up the food in the mouth.

As we shall discuss in Chapter 6, the accessory jaw bones in the earliest mammals, as in the cynodonts, were concerned with the conduction of sound to the inner ear. Allin (1975) suggested that it would facilitate this conduction if the accessory jaw bones were freed from sutural connection with the dentary. This is clearly true: the selection pressures to improve both the dental and auditory function worked together to produce the lower jaw found in cynodonts and early mammals.

In the earliest mammals, *Morganucodon* (Figures 4.9 and 5.11) and its allies representing the first atherian mammals, and *Kuehneotherium* (Figure 4.13) representing the first of the Theria, the thrust bearing on the end of the dentary has become a definitive condyle, but the reptilian jaw joint between the quadrate and the articular is as well developed, relative to the size of the animal, as it is in *Cynognathus* (Figure 5.10).* Comparing the articular surfaces of the two joints in *Morganucodon*, the contrast between the well-formed, steep curves of the quadrate-articular and the almost flat surfaces of the dentary condyle and the glenoid is particularly striking. There can be little doubt that the reptilian quadrate-articular joint functioned as the hinge to guide the opening and closing of the jaws whereas the mammalian squamosal-dentary joint prevented the dislocation of the jaw by acting as a thrust bearing. This worked much more efficiently than the makeshift measure of the surangular-squamosal contact of some of the cynodonts. It is worth mentioning that some modern insectivores (the soricids) have effectively a double jaw-joint by a division of the normal mammalian articulation. Here the inner joint functions as the hinge and the outer as the thrust bearing, exactly as in *Morganucodon*.

With the evolution of this effective means of preventing the dislocation of the lower jaw under the stresses of mastication, the mammalian cheek-teeth evolved and became more effective. They elongated mesiodistally becoming double rooted and came into a definite occlusal realationship with each other. This is the stage reached by *Morganucodon* and *Kuehneotherium*. The definite occlusal relationship enabled active sharpening or 'thegosis' of the teeth to take place, although this may not have been true of the earliest mammals. In these characters the teeth of mammals differ from those of the mammal-like reptiles.

*In making this comparison that part of the jaw-joint in *Cynognathus* which was between the surangular and the quadrato-jugal on the one hand and the dentary on the other has been excluded. The surangular contact here functioned as a thrust bearing. This function was taken over in *Morganucodon* by the more effective dentary-squamosal joint.

## References

Allin, E. F. (1975) 'Evolution of the Mammalian Middle Ear', *J. Morph., Philadelphia, 147*, 403–37

Bolk, L. (1922) 'Odontological Essays. No. 5. On the Relations Between Reptilian and Mammalian Dentition', *J. Anat. Lond., 57*, 55–75

Crompton, A. W. (1963) 'The Evolution of the Mammalian Jaw', *Evolution, Lancaster, Pa., 17*, 431–9

Crompton, A. W. (1972) 'The Evolution of the Jaw Articulation in Cynodonts' in K. A. Joysey & T. S. Kemp (eds.), *Studies in Vertebrate Evolution*, Oliver and Boyd, Edinburgh

Edmund, A. G. (1960) 'Tooth Replacement Phenomena in the Lower Vertebrates', *Contributions No. 52*, Life Sciences Division, Royal Ontario Museum, Toronto

Fourie, S. (1963) 'Tooth Replacement in the Gomphodont Cynodont *Diademodon*', *S. Afr. J. Sci., 59*, 211–13

Hopson, J. A. (1971) 'Postcanine Replacement in the Gomphodont Cynodont *Diademodon*' in D. M. Kermack and K. A. Kermack (eds.) *Early Mammals, Zool. J. Linn. Soc., Lond., 50*, Suppl. 1, 1–21

Kermack, K. A. (1956) 'Tooth Replacement in Mammal-like Reptiles of the Suborders Gorgonopsia and Therocephalia', *Phil. Trans. Roy. Soc.* B, *240*, 95–133

Kermack, K. A., Mussett, F. & Rigney, H. W. (1973) 'The Lower Jaw of *Morganucodon', Zool. J. Linn. Soc., Lond., 53*, 87–175.

Kermack, K. A. Mussett, F. & Rigney, H. W. (1981) 'The Skull of *Morganucodon', Zool. J. Linn. Soc., Lond., 71*, 1–158

Parrington, F. R. (1936) 'On the Tooth Replacement in Theriodont Reptiles', *Phil. Trans. Roy. Soc.* B, *226*, 121–42

Patterson, B. (1956) 'Early Cretaceous Mammals', *Fieldiana, Geol., 13*(1), 1–100

Pycraft, W. P. (1928) *Random Gleanings from Nature's Fields*, Methuen, London

Romer, A. S. (1969) 'Cynodont Reptile with Incipient Mammalian Jaw Articulation', *Science, 166*, 881–2

Romer, A. S. (1970) 'The Chañares (Argentina) Triassic Reptile Fauna. VI. A Chiniquodont Cynodont with an Incipient Squamosal-dentary Jaw Articulation', *Brevoria, 344*, 1–18

Romer, A. S. & Price, L. W. (1940) 'Review of the Pelycosauria', *Geol. Soc. Amer. Sp. Papers* no. 28

Simpson, G. G. (1941) 'The Function of Sabre-like Canines in Carnivorous Mammals', *Amer. Mus. Nov.* no. 1130, 1–12

Watson, D. M. S. (1951) *Paleontology and Modern Biology*, Yale University Press, New Haven

# 6 THE EVOLUTION OF MAMMALIAN SIGHT AND HEARING

## The Competitors of Early Mammals

The Middle Jurassic gives glimpses of the evolving mammals. From recent discoveries (see Chapter 7 for details) there is evidence of a much wider range of mammalian life than was seen in the Upper Triassic and Lower Jurassic, but unfortunately the new material is almost entirely teeth with but a few jaw fragments. It is impossible to reconstruct the skull of a Middle Jurassic mammal in the same way as the Lower Jurassic *Morganucodon*. Nevertheless these later mammals show one quite fundamental difference from the earlier forms: the majority of them have lost the reptilian quadrate jaw articulation. Clearly this fundamental step in the evolution of mammals happened between the Middle and Lower Jurassic, and we must consider why this change came about.

During the Late Carboniferous, Permian and Early Triassic the mammal-like reptiles were the dominant land animals, as are the mammals today; but as the Trias progressed the therapsids came under greater and greater pressure (see Table 1.2). In the Palaeozoic the non-mammal-like reptiles were not a conspicuous part of the fauna; but in the Triassic they began to evolve rapidly and radiated into the ecological niches previously occupied by the mammal-like reptiles (see Table 3.1). Initially it was the primitive archosaurs, grouped together as the Thecodontia, which were the main competitors of the Therapsida. Later it was the descendants of the thecodonts, the advanced archosaurs such as dinosaurs and crocodiles, which caused the final complete collapse of the medium sized and large mammal-like reptiles, so that they vanished completely leaving no descendants. The new competitors of the small mammal-like reptiles belong to another sub-class of the diapsid reptiles — the Lepidosauria. The most primitive order of the Lepidosauria, the Eosuchia, are extinct; but the other two orders are represented in the world today. They are the Rhynchocephalia and the Squamata. The latter are familiar to everyone as they comprise the lizards and the snakes, while the only living species of a rhynchocephalian is the Tuatara (*Sphenodon bipunctatum*), which is confined to a few islands in the Bay of Plenty in New Zealand. As the Squamata were the last to appear, being unknown before the Upper Jurassic, the main competitors of the small mammal-like reptiles and the early mammals in the Upper Triassic and the Lower Jurassic were the Eosuchia and the Rhynchocephalia.

We are mammals and to us it seems natural that the dominant land animals should be mammals; for some reason, possibly arrogance, the birds and insects are usually forgotten. So the sudden collapse of the larger mammal-like reptiles in the Trias can cause some surprise. The new diapsid reptiles replacing them had made fundamental advances in locomotion and hearing so they were really more efficient. The small mammal-like reptiles escaped extinction by becoming nocturnal and so lived on to become the direct ancestors of the mammals.* Up to this time it is probable that there had been no nocturnal tetrapods so the night was 'empty' to be colonised. Present-day mammals are largely nocturnal whilst the majority of living reptiles are diurnal. This difference almost certainly goes back to the end of the Trias or the early Jurassic, when the first mammals became nocturnal to avoid direct competition with the highly successful and efficient diurnal small diapsids.

## Sight in a Nocturnal Environment

The night is a more exacting environment than is the day. With the much lower light intensity, eyes have to become modified to enable them to function. Sight needs to be supplemented by an acute sense of smell and also by acute hearing, perhaps assisted by some form of echo-location or sonar. Finally temperatures are lower at night so there are the problems of maintaining a body temperature high enough for reasonable metabolic activity.

The sensitivity of the eye depends primarily on the light-sensitive cells in the retina. The colour sensitive 'cones' seem to have been the primitive light-sensitive cells of the vertebrate retina. The other type of light-sensitive cells are called 'rods' and are not concerned with colour vision but are much more sensitive to low intensities of light than are the cones. Both types of cell are found in the human retina, which has the cones concentrated towards its centre and the rods at the periphery. For normal light intensities we mainly use the centre of the retina and we can appreciate colour; in these circumstances we rely upon the cones. At low light intensities, a walk in the moonlight for example, we make use of the rods in our retina and colour vision is absent. At very low light intensities, such as trying to detect a faint star in the heavens, it is easier to do so by looking slightly away from its expected position, then the image is received on the periphery of the retina where the rods are most dense.

Most living mammals have retinae which consist predominantly of rods. The only real exception are the Primates, the order that includes man, the monkeys and apes; and this is the only mammalian order whose members are predominantly diurnal. It is an old order, the first primate fossil remains being found in late Cretaceous deposits, but even so the cones were probably secondarily derived from rods. The rod-retinae of the great majority of mammals are a straight heritage from the Mesozoic, when their ancestors become nocturnal to avoid diapsid competition.

*Jerison (1973) gives a good recent summary of the arguments for this belief.

## The Olfactory Sense in the Nocturnal Environment

Most mammals, again with the exception of the diurnal primates, make great use of smell. The well-developed turbinal or 'scroll' bones in mammals are probably more concerned with increasing the area of the olfactory epithelium than acting as a heat-exchanger to warm the air passing into the lungs. The later mammal-like reptiles had already evolved turbinals which their descendants could develop and modify to suit their needs (Chapter 2).

## Hearing in Mammal-like Reptiles

The early mammal-like reptiles lacked an ear-drum; this does not mean that they were deaf only that they were unable to detect air-borne sound. An ear-drum or tympanic membrane is essential if an ear is to have reasonable sensitivity to the minute sound waves propagated through the air. The earliest mammal-like reptiles — the Pelycosauria — could only detect and hear sounds transmitted through the ground. As Tumarkin (1968) demonstrated, they could hear either by what he called the *direct route* or by the *reversed route* (Figure 6.1). In the former, the lower jaw had to be in contact with the ground, so that sound waves were transmitted through the dentary to the other bones of the lower jaw and thence to the articular and quadrate. The stapes always forms an articulation with the quadrate in mammal-like reptiles, so the sound waves continued their journey through the former bone eventually to reach the inner ear, which is housed in the otic capsule firmly attached to the skull. The sound-sensitive cells of the cochlea of the inner ear detect movement in the endolymph or liquid in which they are immersed; they are unaffected by changes in pressure. So in direct-route hearing the vibration of the stapes causing movement in the endolymph enables the animal to hear. The skull with the otic capsules attached forms the fixed reference point and does not vibrate.

In reversed-route hearing the sound was conducted up the animal's legs to the vertebral column, thence to the skull and otic capsule. Thus in this type of hearing the otic capsule itself vibrates, whilst the stapes remains motionless to act as the fixed point required for the endolymph movement to be passed to the cochlea cells. In the pelycosaurs, the stapes is always a remarkably massive bone. In a large *Dimetrodon* (Figure 1.8), the stapes would have been about the thickness of the human thumb. How so massive a bone could function as part of the hearing apparatus of early mammal-like reptiles has been a matter of some surprise. This surprise, that *Dimetrodon* and the many forms with a similar auditory apparatus could hear, meant that reversed-route hearing was not understood. Most palaeontologists persisted in thinking in terms of a stapes transmitting sound from a vibrating tympanic membrane as in our own ears. This theory was first put forward by Broom

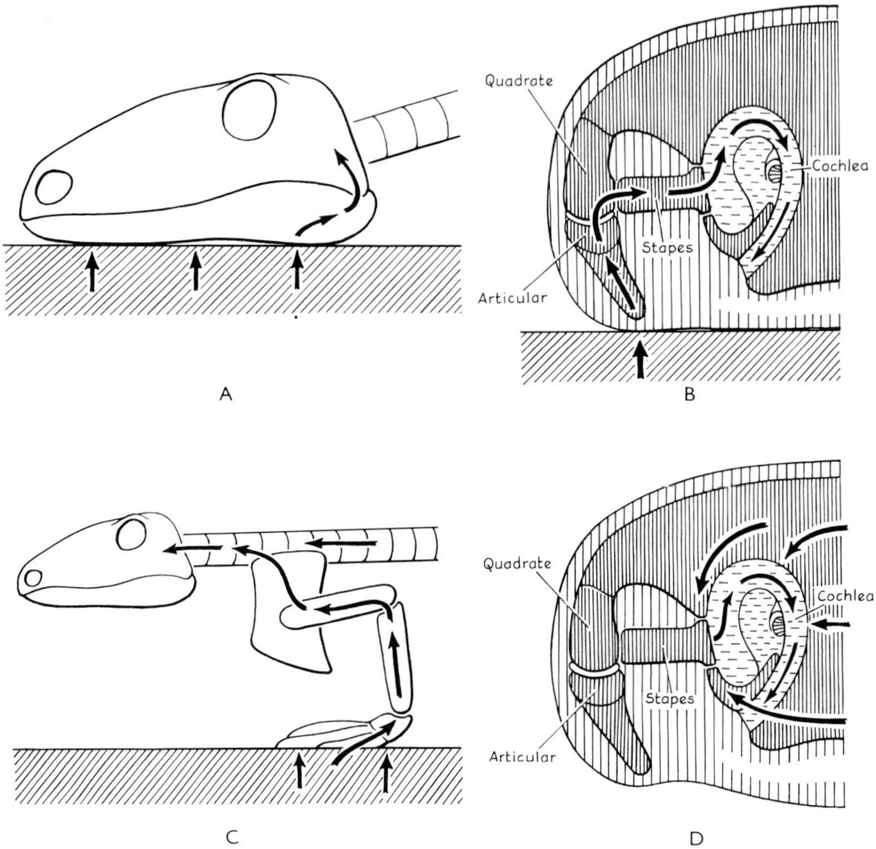


**Figure 6.1:** A and B, *Direct-route Hearing*; Vibrations of the Stapes Set the Endolymph in Motion, the Otic Capsule being the Fixed Reference Body. C and D, *reverse-route hearing*; vibrations of the otic capsule set the endolymph in motion, the stapes being the fixed reference body

(1912, 1936). Parrington (1979) gives the most modern summary of it. So obsessive was this concept that there were even well-meaning, but utterly misguided attempts to give *Dimetrodon* an ear-drum, although Watson (1943) had shown beyond all reasonable doubt that it was impossible. In *Dimetrodon*, the stapes conducted sound waves in a way analogous to that in which electricity is conducted through a wire. The larger the cross-sectional area of the wire the lower its resistance; similarly the large stapes of *Dimetrodon* would conduct sound waves more efficiently than one of smaller cross-section. As Tumarkin (1968) showed, in reverse-route hearing the stapes acts as the fixed reference body while the otic capsule vibrates; so the more massive the stapes the better it functions as the reference body. Tumarkin quotes an experiment on reversed-route hearing in man. If a tuning-fork is placed on the top of the subject's head he hears the note, his auditory ossicles and tympanum acting as the fixed reference body. Now, if the external auditory meatus (external ear passage) on one side of the head is filled with mercury, he hears the sound much more loudly. The mercury increases the effective mass of the reference body. The massive stapes of *Dimetrodon* was well suited to act as such a reference body and it was made still more effective by its abutment against the quadrate thus enabling the mass of the lower jaw to make its contribution. So the ear of *Dimetrodon* was well designed to hear ground-borne sound but there were no means of perceiving purely air-borne sounds.

## The Location of the Tympanic Membrane

In the later Permian and in the Triassic the mammal-like reptiles became more active and agile but at the same time they came under increasing competition from the evolving diapsid reptiles. The latter had tympanic membranes or ear-drums and consequently possessed ears which were efficient receptors of air-borne sounds. For a long time these later mammal-like reptiles, the Therapsida, were considered to have had a tympanum placed directly behind the quadrate, in much the same position as it occupies in the sauropsid reptiles. In 1975, however, Allin showed quite clearly and definitely that a tympanum in this position just could not work in the therapsids. The tympanum would have been too small to be of any use. Allin went back to the old idea, suggested earlier by Watson amongst others, that the tympanum of therapsids was in the lower jaw. One of the bones of the lower jaw, the angular, has a ventrally directed process (reflected lamina) which could have supported a tympanic membrane of a reasonable size. Watson (1951, 1953) considered that a tympanic membrane could have been so supported in the Therapsida and Allin came to the same conclusion. He worked out its implications more completely and in greater detail. Figure 6.2 shows *Thrinaxodon* with the tympanic membrane restored following Allin. Kermack *et al.* (1981) take the analysis further, treating the acoustical problem mathematically. The problem is considered further by Kermack (1983) and Kermack and Mussett (1983).

*Morganucodon* and *Kuehneotherium*, which retained the reptilian jaw articulation, must have had the tympanic membrane in the lower jaw, exactly like the therapsids (Kermack *et al.*, 1981). Figure 4.9 shows the situation of the tympanic membrane in *Morganucodon*.

The lower jaw may seem a peculiar place to house an ear-drum, but one family of present-day lizards do just this. The Amphisbaenidae are a rather peculiar and isolated group of legless and burrowing lizards, which have a sort of analogue of a tympanic membrane formed by the modification of part of the skin of the outer surface of the lower jaw; a long process on the stapes connects this tympanum to the inner ear. This ear works quite well as a

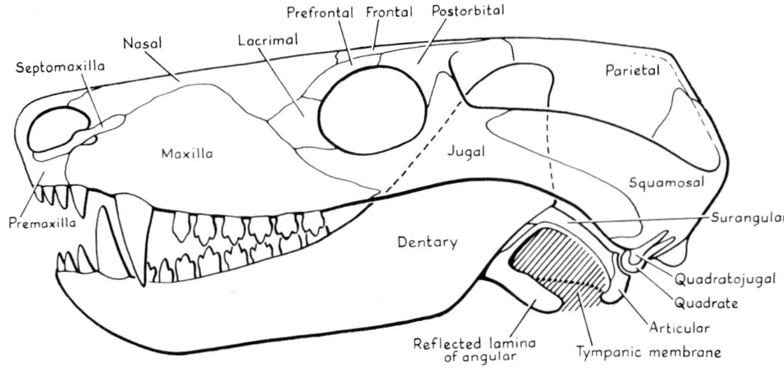


**Figure 6.2:** The Skull of the Cynodont *Thrinaxodon*, with the Tympanic Membrane Restored Following Allin (1975)

detector of air-borne sounds, but it has one major disadvantage, namely it has a very poor response to the higher frequencies. Up to about 1000 cycles per second (1 kHz) it is reasonably good, but above this frequency its efficiency falls markedly and quickly (Wever, 1978). This failure is quite simply due to the excessive mass of the stapes: a stapes long enough to connect the tympanum half-way down the lower jaw to the inner ear in the otic capsule attached to the skull is necessarily massive.

In the therapsids the sound was not transmitted directly by the stapes from the tympanum in the lower jaw to the inner ear. It passed through the bones at the posterior end of the lower jaw, the articular, to the quadrate and thence to the stapes (Figure 6.3). These facts supply one reason, but not the only one, for the posterior bones of the lower jaw of a cynodont forming a loose, and not a bony sutural, contact with the dentary, the largest bone of the lower jaw. They also explain why the quadrate in Therapsida is never fused to the squamosal, but always lies free in a recess in this bone, which supports it. Clearly, if the quadrate is to be part of a chain of bones transmitting sound from the tympanic membrane in the lower jaw to the inner ear, the quadrate must be capable of at least slight movement relative to the skull.

For a tetrapod to perceive air-borne sounds, a tympanum of adequate size and an efficient means of transmitting the sound vibrations from it to the inner ear are essential (Kermack and Mussett, 1983). The bones or auditory ossicles, which transmit the sound vibrations must be as light and free from restraint as possible if there is to be a good response to the higher frequencies. The analogy with the pick-up head of a gramophone is obvious. In cynodonts or in the primitive mammals, such as *Morganucodon*, while there was a tympanic membrane of adequate size, the chain of bones between the tympanum and the inner ear were anything but light and free from restraint. It was made up of the articular, quadrate and stapes. These were comparatively massive bones, and as in the amphisbaenid lizards, the high frequency response must have been poor.

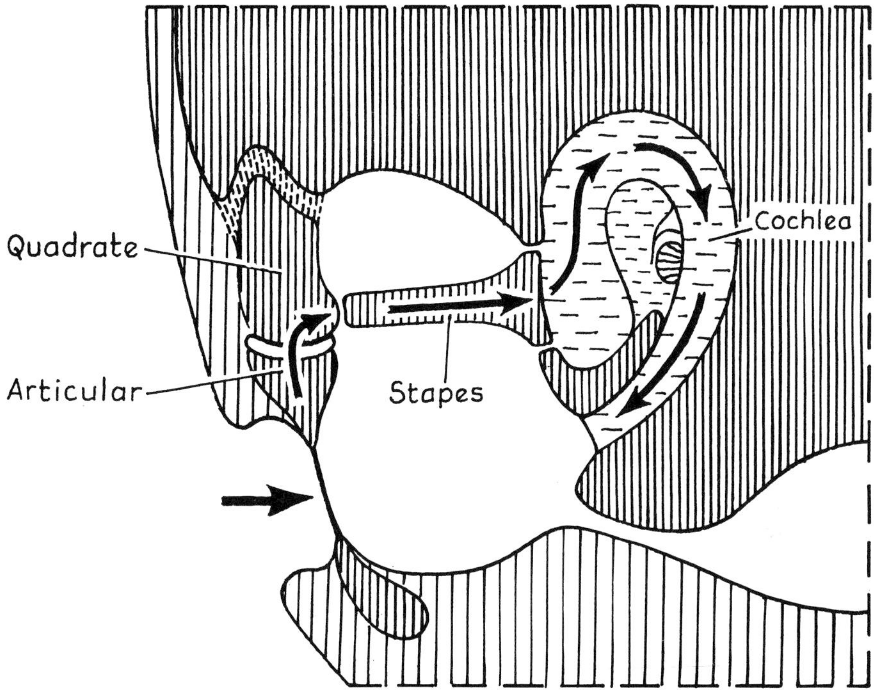


**Figure 6.3:** Diagram to Show How the Ear of a Cynodont Functioned. Air-borne sound vibrations, coming in the direction marked by the arrow, impinge on the tympanic membrane. The membrane vibrates so setting the light articular, quadrate and stapes in motion. The whole complex functions as the necessary acoustic transformer, and sets the endolymph of the cochlea in motion

## Location by Sound

The lepidosaurs had become serious competitors of the first mammals and they could have had excellent high-frequency hearing; this is indicated by living geckos, whose ears are every bit as good in this respect as those of modern mammals. Good reception at the higher frequencies is particularly necessary to the nocturnal animal and this was the mode of life, which had been forced upon the early mammals by lepidosaur competition. A nocturnal animal is more dependent upon its ears for long-range information than is a diurnal one. Eyes, even if modified by a rod-retina backed by a reflecting layer or tapetum are inevitably much less efficient by night than by day. For two reasons high frequencies are best for obtaining the accurate location of objects by night. Firstly an animal locates the direction of the source of a sound by comparing its loudness as perceived by both ears and then by appreciating the time difference between the reception of the sound by the two ears. Secondly, high frequency sounds are not diffracted around obstacles but travel in straight lines.

Considering the first reason in greater detail, for the accurate location of objects at night to be efficient the wavelength of sound must be of the same order as the size of the head, otherwise the sound waves would simply diffract round the skull, as the waves on the sea diffract round a buoy. If this happens the ear on the side of the head away from the source of the sound will hear as loudly as the ear directly facing the source. To make an effective 'shadow' the dimensions of the obstacle must be at least a quarter of a wavelength. At 20,000 Hz and comfortably within the frequency range accessible to a domestic cat, the wavelength is just under 17 mm. At this frequency the head of even a small animal will act as an effective obstacle, putting in 'shadow' the ear facing away from the source of the sound. The animal will be able to locate the sound source with accuracy by turning its head until both ears hear the sound equally loudly. We do not know if Jurassic mammals had external ear pinnae like those of modern mammals. Such a pinna is a highly efficient directional collector and turning the pinnae can facilitate the accurate location of the source of the sound. As everyone knows a cat makes great use of its mobile pinnae in this way. But there is a catch; again the dimensions of the ear pinnae must be of at least the same order of wavelength for them to function as directional collectors. Hence, particularly in small animals, high sound frequencies are necessary.

The second reason for the necessity to utilise the higher frequencies is related to the first. Low frequency sounds may be diffracted around fairly large obstacles (261 Hz is about the frequency corresponding to middle C and the corresponding wavelength is 1.27 m) whilst high frequency sounds travel in straight lines. Early Jurassic mammals were tiny, so that the radius over which they would need to receive information would be correspondingly small, as would be the size of objects of significance to them. Only by the use of high frequencies could they obtain the information they needed.

So far we have discussed the use of hearing in locating prey and predators at night in the context of the predator receiving noises made by its prey, interpreting them and thereby obtaining the bearing of the source. This may be called *passive homing* by analogy with modern missile technology. Some living small mammals employ *active homing* by making use of sonar. Bats (Microchiroptera) for example, emit short ultrasonic pulses which, after reflection from the prey or an obstacle, are heard by the bat giving it the bearing of the object. By measuring the time lapse between emitting and receiving the pulse, the bat can also estimate the range of the reflecting object. This is exactly analogous with radar, using radio waves in air, or sonar, using high frequency sound waves in water. This is why we call it sonar in mammals, although it must be remembered that mammals use the principle in air as well as in water.

Air-borne sonar, as used by a bat, employs frequencies ranging from about 13,000 Hz to more than 100,000 Hz with pulses ranging in duration from one to ten milliseconds. Since a one millisecond pulse comprises a chain of sound waves some 330 mm in length, short pulses are essential if the bat is to be able to detect objects close to it. These short pulses demand a high frequency source, since a low frequency wave chopped in this way would become simply a series of wide-spectrum clicks. In addition the prey must be a good reflector of sound pulses if it is to be detected. Bats employing sonar feed on flying insects. For a sound wave to be reflected the dimensions of the object must be at least of the order of half a wavelength. 13,000 Hz corresponds to a wavelength of 25 mm and 100,000 Hz to a wavelength of 3 mm, so using these frequencies bats can detect objects ranging from 12 mm to 1.5 mm, which covers the size of flying insects very nicely.

*Active homing* on prey is found in primitive insectivores and is probably to be found in other primitive mammals. It is also employed in water by whales and dolphins (Cetacea). The shape of a natural cast of the brain of a primitive Palaeocene carnivore *Miacis* indicates that it used active homing or air-borne sonar. Indeed sonar of this kind may have evolved in the Jurassic when mammals first took to a nocturnal mode of life. If this be so it would supply another powerful reason for the early mammals to improve drastically the efficiency of their ears as receptors of high frequency sounds. It may be asked, if sonar was evolved so early in the history of the Mammalia, why do so many of the more highly evolved mammals, such as the fissipedes and primates, lack it. The reason may well be that the pulses sent out by the animal employing sonar can alert the recipient, be it prey or predator, to the presence of the former. Some night-flying moths have developed receptors to detect bat sonar, which enables them to take appropriate evasive action. There is a human analogy here. In the last war the British heavy bombers used a side-looking radar (H2S) for navigation. The Germans were able to make use of the radio waves emitted by a bomber to 'home in' fighters on to it. At the present time there are guided missiles which 'home in' on enemy radar stations using the same principle. These examples in terms of animal ecology, are a predator 'homing in' on a prey that is transmitting high frequency sounds, or a prey detecting a predator by monitoring the latter's transmissions. A highly evolved predator such as a cat with efficient night vision may do better homing passively on the sound made by its prey, so keeping its victim in ignorance of its danger. Be this as it may, the primitive mammals of the Early Jurassic, with less efficient senses than a modern fissipede, were under great selection pressure to improve the high frequency response of their ears.

## The Evolution of the Mammalian Middle Ear

To obtain the required high frequency response the moving parts, namely the chain of bones which connect the tympanic membrane to the inner ear, had to become lighter and much freer from constraint. The chain of bones consisted of the articular, quadrate and stapes and as long as this had a dual function of conducting sound and of forming the jaw articulation it was impossible to achieve the necessary reduction in mass and freedom from constraint. To do this reptilian jaw articulation had to disappear so that the quadrate and articular could become incus and malleus and solely auditory in function. The retroarticular process becomes the manubrium of the malleus which contacts the tympanum. The early mammals then relied on the squamosal-dentary articulation to form the jaw-joint. This was perfectly satisfactory: all modern mammals depend upon the articulation of these two bones for their jaw-joint or hinge. The measure of the change thereby made possible can be seen by comparing the quadrate and articular of a cynodont such as *Cynognathus* (Figures 4.5 and 5.10) with the homologous bones (incus and malleus) of a modern mammal with a similar sized skull, a pig being a suitable example. In the former these bones are massive and are firmly bonded by ligaments to neighbouring bones, whilst in the latter they are represented by tiny fragile auditory ossicles joined together by small thin ligaments to form a 'chain with moveable links' across the middle ear. When the articular and quadrate became the malleus and incus respectively, the angular retained its function of supporting the tympanic membrane and became the tympanic bone.

The change described took place quickly between the Lower Jurassic, when all the fossil mammals we know have a double jaw articulation, and the Middle Jurassic when all forms except the docodonts have a fully mammalian jaw articulation. The loss of the quadrate-articular jaw hinge is quite certain; the trough on the medial surface of the dentary, which in *Morganucodon* (Figures 4.9 and 5.11) houses the articular, prearticular, angular and surangular is completely absent in all Middle Jurassic mammals except the Docodonta. That the mammals which had lost the reptilian jaw-joint had also acquired the mammalian middle ear is inference. The tiny auditory ossicles are so loosely attached that there is little chance that they will ever be found *in situ* in the skull of a Middle Jurassic mammal. Nevertheless it is a perfectly reasonably inference. This is indicated by the condition of the jaw-joint and ear ossicles in the two lineages of modern mammals, namely the Atheria or monotremes and the Theria representing all other living mammals. These two lineages have been separate since the beginning of the Jurassic at the latest and yet in both the reptilian jaw-joint has been lost and the quadrate and articular have become auditory ossicles. Indeed it is hard to imagine any other reason why the reptilian jaw-joint should have been lost.

The exact details are unknown as to how the change took place, but a similar change does take place in the embryonic development of living marsupials. They are born in a very early stage of embryonic development and the newly born young 'swims' actively with its fore-limbs from the mother's vulva to the pouch. Here the little embryo opens its mouth and seizes a teat and once held, the teat is not relinquished until the young animal, no longer an embryo, can leave the pouch. Milk is pumped down its throat by the muscles of the mother's udder, whilst it remains passively attached to the teat. When it is first born the embryo has its functional jaw-joint between the articular and quadrate and this reptilian joint functions when the mouth opens initially to take hold of the teat (Palmer, 1913). At this stage the angular is still a bone of the lower jaw but the latter is reconstructed whilst the embryo is attached to the teat. The quadrate and articular pass into the middle ear to become the incus and malleus respectively, whilst the angular becomes the tympanic, a bone of the ear supporting the eardrum. When the young marsupial becomes detached from the teat and opens its mouth it is by a fully mammalian squamosal-dentary articulation.

## References

Allin, E. F. (1975) 'Evolution of the Mammalian Middle Ear', *J. Morph., Philadelphia 147*, 403–37

Broom, R. (1912) 'On the Structure of the Internal Ear and the Relations of the Basicranial Nerves in *Dicynodon*, and on the Homology of the Mammalian Auditory Ossicles', *Proc. Zool Soc., Lond.*, pp. 419–25

Broom, R. (1936) 'On the Structure of the Skull in the Mammal-like Reptiles of the Suborder Therocephalia', *Phil. Trans. Roy. Soc., Lond.*, (B), *226*, 1–42

Jerison, H. J. (1973) *Evolution of the Brain and Intelligence*, Academic Press, New York

Kermack, K. A. (1983) 'The Ear in the Theropsida', *Geobios* mémoire spécial 6, 151–6

Kermack, K. A. & Mussett, F. (1983) 'The Ear in Mammal-like Reptiles and Early Mammals', *Acta Palaeont. Polonica, Warsaw 28* 147–58

Kermack, K. A., Mussett, F. & Rigney, H. W. (1981) 'The Skull of *Morganucodon', Zool. J. Linn. Soc., Lond., 71*, 1–158

Palmer, R. W. (1913) 'Note on the Lower Jaw and Ear Ossicles of a Foetal *Perameles', Anat. Anz. Jena, 43*, 510–15

Parrington, F. R (1979) 'The Evolution of the Mammalian Middle and Outer Ears: a Personal Review', *Biol. Rev., Cambridge, 54*, 369–87

Tumarkin, A. (1968) 'Evolution of the Auditory Conducting Apparatus in Terrestrial Vertebrates' in A. S. de Reuck & J. Knight (eds.), *Ciba Foundation Symposium on Hearing Mechanisms in Vertebrates*, Churchill, London pp. 18–37

Watson, D. M. S. (1943) 'On Permian and Triassic Tetrapods', *Geol. Mag., 79*, 81–116

Watson, D. M. S. (1951) *Paleontology and Modern Biology*', Yale University Press, New Haven

Watson, D. M. S. (1953) 'The Evolution of the Mammalian Ear', *Evolution, Lancaster, Pa, 7*, 159–77

Wever, E. G. (1978) *The Reptile Ear*, Princeton University Press, Princeton

# 7 THE COMPLETION OF THE FIRST STAGE OF MAMMALIAN EVOLUTION IN THE MIDDLE JURASSIC

## Stonesfield Mammals

The existence of mammals in the Middle Jurassic has been known since the eighteenth century when they were first discovered in the Stonesfield Slate of Oxfordshire. The first specimen was thought not to be mammalian, and the first recognition that mammals already existed in the Mesozoic Era did not happen until a century later. A young man, William John Broderip, studying Law at Oxford, was interested in Natural History and attended lectures by William Buckland, later elected to the first professorial Chair of Geology at Oxford in 1819. In 1812 or 1814, the date is uncertain, Broderip came to Buckland with two tiny lower jaws which he had obtained from an old quarryman, who was collecting for him at Stonesfield, a few miles north of Oxford. The Stonesfield Slate is a limestone which, on exposure to the weather and particularly to frost, splits easily into thin lamellae. It was used, and a similar stone is still used, in the district as roofing slates giving the houses and farms their attractive and regionally characteristic appearance. The Slate was quarried, or with more accuracy mined, at Stonesfield, the blocks being stacked for weathering to take place and then finally split by hand into roofing slates. Wages were low and the quarrymen knew that some of the gentlemen of Oxford would pay good money for interesting fossils. This is how Broderip obtained his two jaws and also how most of the Stonesfield mammals were collected.*

Both of Broderip's specimens showed that the lower jaw consisted of a single bone, the dentary with a condyle for articulation with the skull. Broderip and Buckland consequently saw quite correctly that the jaws were mammalian. At that time no pre-Tertiary mammals were known, and it was universally accepted that the Mammalia did not exist in the Mesozoic. In consequence Broderip and Buckland refrained from publishing their discovery even though the great French palaeontologist Cuvier visited Oxford in 1818, saw the specimens and confirmed their mammalian nature. In 1823 Buckland finally published a brief account of the specimens, quoting Cuvier's authority for their mammalian nature. In the next year Prevost published the first detailed description including a figure of one of the two specimens, which Buckland had purchased from Broderip sometime before. A controversy then arose as to whether or not the specimens were mammals and the matter was not finally resolved until Sir Richard Owen published on

*Simpson (1928) gives a full account of the history of these discoveries.

the specimens in 1838 and 1842. Since then their mammalian nature has never been doubted.

There is a total of eleven specimens of mammals from Stonesfield and also two specimens of *Stereognathus*. This is the last of the tritylodonts. The workings at Stonesfield have long been abandoned, so despite there being extensive tips, which have been worked over by collectors looking mainly for invertebrate fossils, it is unlikely that any further mammalian material will be obtained from the site. The last specimen was discovered over a hundred years ago and so we shall have to make do with what we have. The eleven specimens are included in two separate orders, the Triconodonta and the Pantotheria (Table 3.1). They are all lower jaws. Four of them are of the pantothere *Amphitherium prevostii*, whilst the remaining seven are

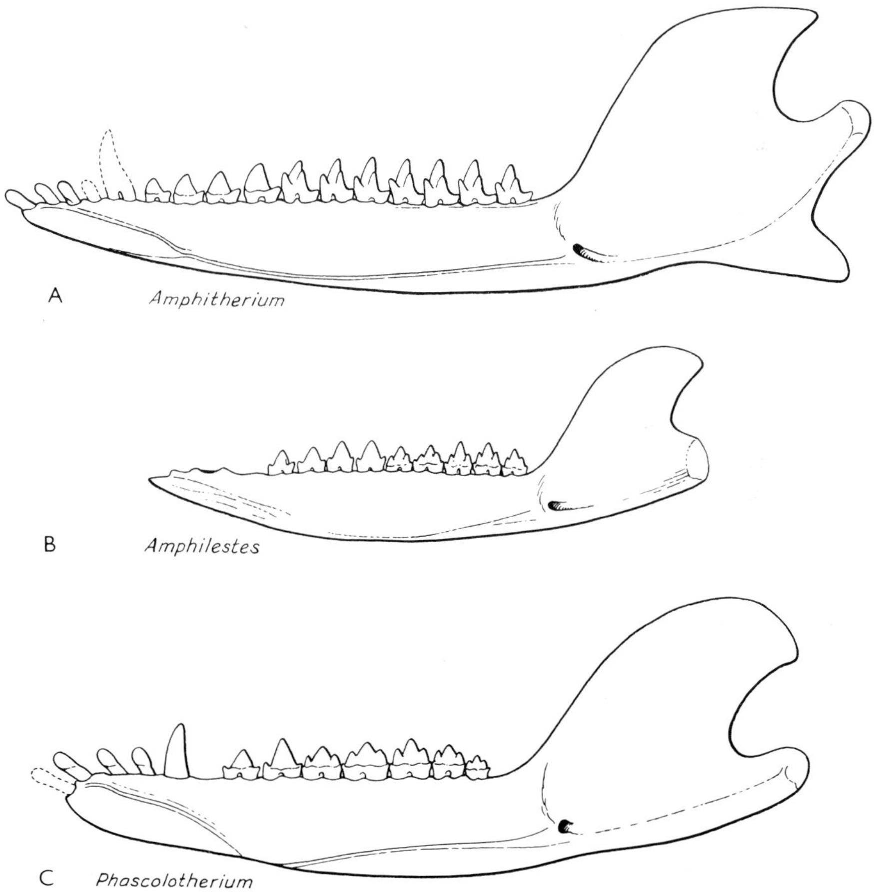


**Figure 7.1:** The Lower Jaws of the Stonesfield Mammals Seen in Medial View. Notice the absence of the trough for the accessory jaw bones in the dentary. Compare with the lower jaws of *Morganucodon* (Figures 4.9 and 5.11) and *Kuehneotherium* (Figure 4.13)
Source: After Simpson, 1928

triconodonts belonging to two separate genera and species, *Amphilestes broderipii* and *Phascolotherium bucklandi* (Figure 7.1). It is a happy chance that the specific names commemorate the three men who first recognised the mammalian nature of Broderip's fossils. It should be recalled that in the Lower Jurassic, particularly in the Welsh fissures there is a similar association of a triconodont, *Morganucodon* and a pantothere *Kuehneotherium*. But there is a fundamental difference between the Lower Jurassic and Middle Jurassic mammals. The jaw-joint in the Stonesfield mammals was solely between the dentary and the squamosal and in this respect these animals were fully mammalian. These Middle Jurassic mammals lacked the groove at the back of the dentary, which in *Morganucodon* and *Kuehneotherium* housed the articular and angular along with other accessory jaw bones (see Chapter 4). The missing bones have migrated to the middle ear region, so the double jaw articulation and associated middle ear of a strictly cynodont pattern, characteristic of the Lower Jurassic mammals, has been lost. It is improbable that the lower jaw in the Stonesfield mammals consisted only of a single bone, the dentary; it is possible that two bones of the reptilian lower jaw, the splenial and coronoid, were still present in the living animal, but being tiny relic bones they did not survive post-mortem and fossilisation hazards. It should be emphasised that not all Middle Jurassic mammals had lost the reptilian jaw articulation, only that the mammals found in the Stonesfield Slate deposits have done so.

The mammals from Stonesfield are divided into two clearly distinct groups, one genus, *Amphitherium*, representing the Theria, and the other two genera *Amphilestes* and *Phascolotherium*, representing the Atheria (Figure 7.1). The Jurassic Theria are grouped in an infra-class of their own, the Pantotheria. Until the discovery of *Kuehneotherium* about thirty years ago, *Amphitherium* was the oldest known member of the Theria. Although all that is known of the latter is the lower jaw and the lower dentition, the early discovery of *Amphitherium* made it inevitably the focus of all hypotheses of the origin of the dentition of the modern Eutheria. Although it is certainly not ancestral to any of the living mammals, its dentition is nevertheless of great interest (Mills, 1964, 1971).

Considering pantotherian dentitions generally, it will be observed that they contain more post-canine teeth than are found in the modern Eutheria. Typically, marsupials do not have more than three premolars and four molars, while placentals have primitively four premolars and three molars. Thus in both cases there are seven post-canines in all. In *Amphitherium* there were most probably eleven post-canines, consisting of four premolariform and seven molariform teeth. The tendency to have more than seven post-canines is common to many pantotheres and we have no idea of its significance. The molars of placental mammals are never replaced. Unfortunately we do not know if the molariform teeth of *Amphitherium* were true molars in this sense.

Comparing the teeth of *Amphitherium* from the Middle Jurassic with the those of the Lower Jurassic *Kuehneotherium* (Figure 4.11), the lower molars in both animals have a large trigonid which is much larger in the former than in the latter. The molars of *Amphitherium* have, for a Jurassic mammal, a large talonid, but it is in no way basined. In both mammals the paracone of the upper molars sheared across the buccal face of the talonid, making a great wear facet. In *Kuehneotherium* many specimens show this facet, and Simpson (1928) figures it clearly in *Amphitherium*.

*Amphitherium* has always been accepted, perhaps too readily, as ancestral to modern mammals; but the position of the other two genera from Stonesfield — *Amphilestes* and *Phascolotherium* (Figure 7.1) — is more debatable, despite one specimen of the former being the first Mesozoic mammal ever to be discovered.*

*Amphilestes*, like *Amphitherium*, is known only by the lower jaw and dentition, and in both the complete lower jaw is about 25 mm long. That of *Phascolotherium* was a little larger, about 35 mm. In the two Atheria there is a well-developed condyle on the dentary and there is no sign of the trough for the jaw bones which form the second or reptilian jaw-articulation in the Lower Jurassic atherian, *Morganucodon* (Figure 4.9). We have discussed earlier (Chapters 5 and 6) the selection pressures due to competition with diapsid reptiles, causing the loss of the reptilian jaw articulation and the incorporation of the bones involved in the mechanism of the ear.

In the molar teeth of *Amphilestes* and *Phascolotherium* there are three cusps arranged in a row along the axis of the jaw in each tooth, the centre cusp being the largest. In *Amphilestes* the central cusp is relatively higher and more slender than in *Phascolotherium* but the fundamental pattern is the same in both. Although the teeth are referred to as 'molar' it should not be assumed that they are replaced once as are the molar teeth of placentals. It is not known how many times the molar teeth of *Amphilestes* were replaced or even if they were replaced at all. Unfortunately if all there is of an animal is the lower jaw the animal has to be classified largely upon the character of the teeth. This is not always the most satisfactory procedure, particularly when, as here, the teeth in question do not really closely resemble those of any better known group of mammals. The teeth of *Amphilestes* and *Phascolotherium* from the Stonesfield Slate most closely resemble those of two equally obscure and little known genera *Phascolodon* and *Aploconodon* from the Morrison Formation (Uppermost Jurassic) of the United States of America (Simpson, 1929). Whether these resemblances indicate relationship we do not know. Even if there were a relationship, it would not be of much help as the taxonomic position of the American mammals is unknown.**

The teeth of the two Stonesfield triconodonts are usually compared with

*Sir Christopher Sykes possessed a lower jaw of *Amphilestes* in 1764.

**Cassiliano and Clemens (1979) consider *Phascolodon* and *Aploconodon* as symmetrodonts.

those of the Upper Jurassic triconodontines (Chapter 8); and they may also be compared with those of *Morganucodon*. All have in common a linear arrangement of the cusps in the molars as against the triangular arrangement found primitively in the Theria, and illustrated by the teeth of *Kuehneotherium* (Figures 4.10 and 4.11) and *Amphitherium* (Figure 7.1). It is this linear arrangement of the cusps which justifies the placing of *Amphilestes* and *Phascolotherium* in the order Triconodonta. This is the classification that is usually adopted but we must keep clearly in mind that the evidence is not strong. Mills (1971) has suggested that the two Stonesfield forms are not triconodonts at all, but aberrant Theria. He comments on resemblances between the teeth of *Amphilestes* and *Phascolotherium* and some atypical molars of *Kuehneotherium* and he may well be right.

## Other Middle Jurassic Mammals

Recently, two other formations have yielded Middle Jurassic mammals. Only preliminary acccounts have so far been published but it is already clear that mammalian life in the Middle Jurassic was much richer and more varied than ever imagined. As a corollary the evolutionary significance of the Stonesfield mammals, and *Amphitherium* in particular, may have been overestimated in the past. Time, and the full publication of new discoveries will tell.

The first of the new discoveries was made in August 1971 in Skye by Dr M. Waldman and Mr M. J. Dobinson. It is the first, and to date the only, locality in Scotland which has yielded Mesozoic mammals, and the second locality anywhere in the world to be discovered in the Middle Jurassic since the first Mesozoic mammal came from the Middle Jurassic Stonesfield Slate some 200 years ago. So far only preliminary accounts of the finds have been published (Waldman and Savage, 1972); but the fauna, which contains both a docodont and the tritylodont *Stereognathus*, is different from Stonesfield. It is not profitable to discuss this fauna further until more has been published on it.

The other new Middle Jurassic fauna of mammals comes from the Forest Marble, which lies immediately above the Stonesfield Slate in the geological succession, although both are of Bathonian age. The Forest Marble is a marine deposit, and the mammals come from anomalous sediments suggestive of deposition close inshore. An amateur, Mr E. F. Freeman (1976a, b, 1979), first discovered the existence of mammals in the Forest Marble. Both Freeman and University College, London have made considerable collections from this deposit, and, although most of the mammalian fossils are isolated teeth, lower jaws lacking teeth are not uncommon and there is at least one well-preserved fragment of a lower jaw with teeth. Like the Scottish finds, little can be said here until the mammals of the Forest Marble are described more fully. The Forest Marble fauna is remarkable for its variety and may contain the earliest multituberculate.

It is now clear that the Mammalia had undergone a major adaptive radiation by Middle Jurassic times and formed a significant part of the terrestrial vertebrate fauna. The corollary to this is that the significance of *Amphitherium*, until 40 years ago much the earliest therian mammal, may well have been greatly overestimated. The assumption that the collections that are available in museums give a representative sample of the fauna of a particular geological age is a highly dangerous one, although it can be difficult to resist.

## References

Cassiliano, M. L. & Clemens, W. A. (1979) 'Symmetrodonta' in J. A. Lillegraven, Z. Kielan-Jaworowska & W. A. Clemens (eds.), *Mesozoic Mammals*, University of California Press, Berkeley

Freeman, E. F. (1976a) 'A Mammalian Fossil from the Forest Marble (Middle Jurassic) of Dorset', *Proc. Geol. Ass., 87*, 231–6

Freeman, E. F. (1976b) 'Mammal Teeth from the Forest Marble (Middle Jurassic) of Oxfordshire, England', *Science, 194*, 1053–5

Freeman, E. F. (1979) 'A Middle Jurassic Mammal Bed from Oxfordshire', *Palaeontology, 22*, 135–66

Mills, J. R. E. (1964) 'The Dentitions of *Peramus* and *Amphitherium', Proc. Linn. Soc. Lond., 175*, 117–33

Mills, J. R. E. (1971) 'The Dentition of *Morganucodon*', in D. M. Kermack & K. A. Kermack (eds.), *Early Mammals, Zool. J. Linn. Soc. Lond., 50*, suppl. 1, 29–63

Simpson, G. G. (1928) *A Catalogue of the Mesozoic Mammalia in the Geological Department of the British Museum*, British Museum (Natural History), London

Simpson, G. G. (1929) *American Mesozoic Mammalia*, Yale University Press, New Haven

Waldman, M. & Savage, R. J. G. (1972) 'The First Jurassic Mammal from Scotland', *J. Geol. Soc. Lond., 128*, 119–25

# 8 THE MAMMALS OF THE UPPER JURASSIC

## The Durdlestone (Durlston) Bay Mammals

There are three major mammalian faunas in the Upper Jurassic (Tables 3.1 and 8.1). Two have been known for a long time, whilst the third was discovered in the late 1950s. The first comes from Durdlestone (Durlston) Bay, near Swanage in the Isle of Purbeck, Dorset. It is of uppermost Jurassic age (Clemens *et al.*, 1979). Its discovery was made in or just before 1854 as in that year the Rev. P. B. Brodie and a Mr Wilcox sent to Professor Richard Owen some mammal jaws along with other vertebrate fossils that they had collected from Durdlestone. At the same time, according to contemporary accounts, Mr S. H. Beckles carried out further excavations at another site at the top of the cliff at Durdlestone and referred to as 'Beckles' Pit'. From it came a large collection of mammalian specimens, and this can still be considered the most important single collection of Mesozoic mammals ever made. There is an excellent account of the locality in Simpson (1928a).

Unfortunately there is a difficulty about the Beckles' Collection which should be mentioned. The specimens in it are usually stated to have come from the 'Dirt Bed'. This is a technical term for a dark-coloured bed containing a considerable amount of organic matter; such beds are usually considered to be fossil soils. Today there are two Dirt Beds, not one, exposed at beach level at Durdlestone Bay and both have yielded mammals in recent years. The group from University College, London has made a small collection of mammal teeth by collecting bagfuls of matrix from both the Dirt Beds, taking them back to the laboratory and breaking down the matrix with dilute acetic acid, then washing and sorting the residue. Other workers have obtained specimens by more conventional methods, such as collecting bones and teeth from freshly exposed or weathered surfaces of the Bed. Some of the specimens in the Beckles collection are still embedded in the matrix in which they were originally found and this is a limestone which is nothing like, either in appearance or consistency, the material forming the two Dirt Beds exposed on the shore today. Two of the Purbeck specimens in the British Museum (Natural History) are known to have come from limestone layers. One is the only known specimen of *Trioracodon major* which comes from a layer of hard limestone known as the 'Feather'; and the other is the maxilla of *Trioracodon ferox* which was found in the 'Cap' which lies immediately beneath the former layer. So it seems reasonable to assume that mammals occur sporadically throughout a number of adjacent beds in Durdlestone Bay and that they are certainly not confined to a single Dirt Bed.

The further history of Beckles' Pit is frustrating and in no way helps to

| | Period | | Stage | EUROPE | ASIA | NORTH AMERICA | AFRICA |
|---|---|---|---|---|---|---|---|
| | CRETACEOUS | Upr | Cenomanian | | | | |
| 100 | | | Albian | | | M P T Eutheria | |
| | | | Aptian | | M P T Eutheria | | |
| | | | Barremian | M | | | |
| | | Lower | Hauterivian | | | | |
| | | | Valanginian | M P | | | |
| | | | Ryazanian | | | T | |
| | JURASSIC | Upper | Portlandian | M P D T | Theria? | M P D T | P |
| 150 | | | Kimmeridgian | M P D T | | | |
| | | | Oxfordian | | | | |
| | | | Callovian | | | | |
| | | Middle | Bathonian | M P D T H Morg | | | |
| | | | Bajocian | | | | |
| | | Lower | Toarcian | | | | |
| | | | Pliensbachian | | | | |
| | | | Sinemurian | P H Morg | P-India | | |
| | | | Hettangian | | T? Morg | Morg | Morg |
| | TRIASSIC | Upper | Rhaetian | P H Morg | | | |
| 200 | | | Norian | H | | | |
| | | | Carnian | | | | |

KEY:

| | |
|---|---|
| M | Multituberculates |
| P | Pantotheres |
| D | Docodonts |
| T | Triconodonts |
| H | Haramiyids |
| Morg | Morganucodonts |

**Table 8.1:** The Continental Distribution of Localities of Mesozoic Mammals

resolve the matter. Some twenty-five years after Beckles, the Willett brothers reopened the Pit and carried out extensive excavations, only to obtain one mammal jaw. So either Beckles was working a rich but limited pocket which he completely cleaned out or the Willetts did not find the original site of the Pit. Both hypotheses are equally likely. However Willett and Willett (1881) considered the Pit to be in the Upper Dirt Bed, whereas Simpson (1928a) suggests that it was in the Lower Bed. Willett is more likely to be correct. As the traditional site of the Pit has now been built over and is completely inaccessible, the problem is unlikely to be solved.

Like the Forest Marble Beds in the Middle Jurassic, the Purbeck Beds of the Upper Jurassic are deltaic or estuarine in origin, so as well as mammals they contain small, presumably freshwater, crocodiles and the first indisputable lizards. The fossils represent the remains of the bodies of land vertebrates of the time, which were washed into nearby streams and thence down to the delta or estuary.

## The Mammals of the Morrison Formation

The next of the great mammal localities of the Upper Jurassic to be discovered was in the United States of America. The Morrison Formation is of the same age as the Purbeck and occurs in the west in Colorado, Wyoming and Montana. In 1878 W. H. Reed discovered the first mammals in the Morrison at Como Bluff in West Wyoming (Simpson, 1929). Simpson (1928a) comments how the publication of the definitive monograph on the English Mesozoic mammals by Sir Richard Owen in 1871 marks a very real watershed in the history of the discovery of the Mesozoic Mammalia. Prior to that date, Britain had been the country where almost all the Mesozoic mammals known had been discovered and where they were studied and described. After 1871 very little new material was discovered either in Britain or in Europe for the next seventy years. During this period the major discoveries of Mesozoic mammals were made in the North American continent and the USA was the country where they were studied.

The famous Quarry 9 at Como Bluff was worked extensively after its discovery in 1879 and yielded a collection even larger than that from Beckles' Pit. Simpson (1929) tells us that all the Morrison mammals came from Quarry 9, with the exception of only five specimens. The first was picked up on the surface some way from Quarry 9, another was found in Quarry 11 and three came from Cañon City in Colorado. Until the discovery of the Welsh fissure localities after the Second World War, Quarry 9 held the record for the locality producing the largest number of Mesozoic mammal specimens. The Welsh fissures have produced only two, possibly three, genera of mammals, but thirteen genera and 25 species come from Quarry 9. Other localities yielding Mesozoic mammals are known from the Morrison, all in Colorado. The first to be discovered is Garden Park in Fremont County. This has produced only two species of mammal (Simpson, 1929). Two more localities have been discovered more recently, both in Mesa County (Clemens *et al.*, 1979).

## The Guimarota Mammals

The mammalian faunas from the Purbeck and Morrison bear a remarkable resemblance to each other. This is less surprising if it is recalled that in the Upper Jurassic, Europe and North America were joined to form Laurasia and that the Atlantic Ocean still had to be formed (Figures 8.1 and 8.2). They contrast with the third Upper Jurassic fauna of mammals from the Guimarota in Portugal, which is of an older geological age, coming from the Kimmeridgian (see Table 8.1). While the Purbeck and Morrison faunas have been known for at least a century, that from the Guimarota has been known for only a quarter of that time. It was discovered in 1959 by Professor W. G. Kühne, who first discovered the fissure faunas of the Bristol Channel region (Kühne, 1961, 1968a). Although the individual bones and teeth in fissure material are often exquisitely preserved, the mammal skeletons are almost invariably completely dissociated. Kühne hoped for associated skeletons, and he found them at Guimarota in a disused mine in the Jurassic Brown Coal near Leiria. This locality is not a fissure deposit but a seam in the mine, which is still being worked by Kühne's associates at the Free University in West Berlin. Two other sites were subsequently discovered in Portugal by the group, but the Guimarota site remains by far the most important. Recently the complete skeleton of a pantothere has been discovered at this locality by Henkel and Krebs (1977). This is one of the most important vertebrate fossils found this century and when it is described it should provide much valuable information about the early history of the mammals.

Although the Guimarota fauna is rather different from those from Purbeck and the Morrison, examples from three main groups of mammals are found in all three of these Upper Jurassic deposits. Two of the groups, the Triconodonta and the Multituberculata are members of the subclass Atheria, whilst the third, the Pantotheria is the earliest subdivision of the subclass Theria. The order Triconodonta is composed of three suborders, the Morganucodonta, which is unknown in the Upper Jurassic, and the Docodonta and the Eutriconodonta both of which are represented there. A member of the first suborder, *Morganucodon* could be ancestral to members of the other two suborders. *Amphilestes* and *Phascolotherium* are the first eutriconodonts and the first docodont is from Skye and has yet to be described fully (Chapter 7).

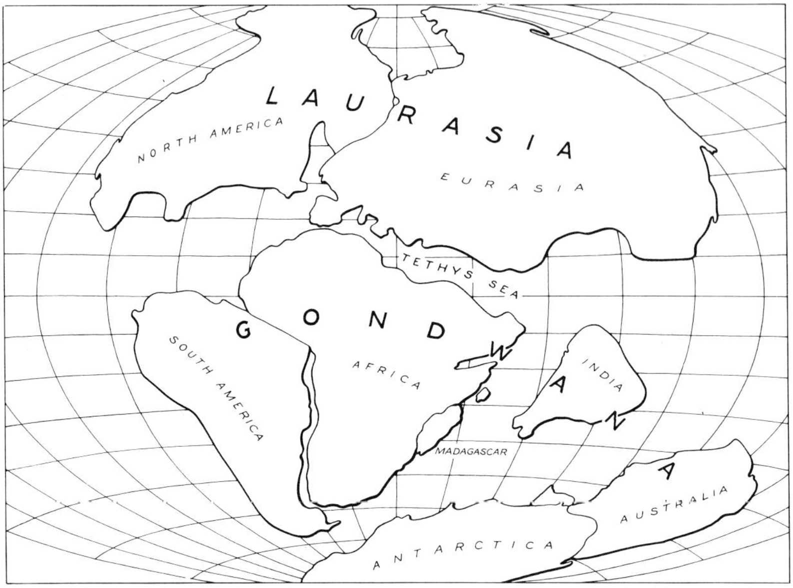


**Figure 8.1:** The World at the Beginning of the Jurassic (135 × $10^6$ years ago), Showing the Beginning of the Opening of the South Atlantic Ocean
Source: After Dietz and Holden, 1970

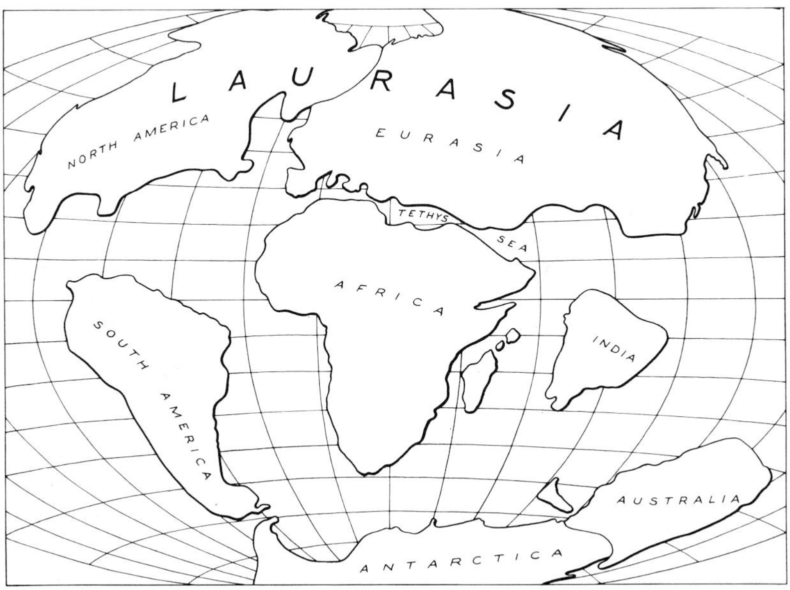


**Figure 8.2:** The World at the End of the Cretaceous (65 × $10^6$ years ago), Showing the Widening of the Atlantic Ocean and the Isolation of South America
Source: After Dietz and Holden, 1970

## Other Upper Jurassic Faunas

Another Upper Jurassic locality, discovered by the German Tendaguru Expedition (1909–1912) is at Tendaguru, in what was formerly German East Africa, and is now Tanzania. Only one mammalian specimen is known from this locality, the lower jaw of a eupantothere, lacking the crowns of the teeth. It has been called *Brancatherulum tendagurense* (Branca, 1916; Dietrich, 1927; Simpson, 1928b).

Recently Chow and Rich (1982) have described a Mesozoic mammal from Sichuan (Szechwan), China. The fossil comes from beds (Upper Shaximiao Formation) which have usually been considered as Upper Jurassic, but recent evidence suggests that they may be older.

The fossil (*Shuotherium dongi*) is a left dentary with some of the cheek-teeth preserved. The remarkable thing about this specimen is that the molars have a pseudo-talonid *anterior* to the trigonid. The authors classify this specimen as an aberrant member of the Theria.

## Triconodonta

The Eutriconodonta and the Docodonta show specialisations in two quite different directions (see p. 109). The former were the small carnivores of their time with modifications to their post-canine teeth which made them highly efficient for slicing flesh and bone. Figures 8.3A and B show a reconstruction of the snout of *Priacodon* from the Morrison. Figure 8.3C shows a reconstruction of the triconodont post-canine dentition. The action of the shearing teeth was quite unlike that of the carnassial teeth of modern fissipedes, in which the molars shear past each other like the blades of a pair of scissors. Many years ago Simpson (1925) showed that in eutriconodonts the crests of the lower molars bit against the upper molars with an action similar to that of some garden pruners, where the sharp blade 'occludes' against an anvil, which in this case represents the upper molars (Figure 8.3D). The action of both the pruners and the eutriconodont teeth is so efficient at slicing up tough material that it can be misinterpreted as a shearing one.

All the eutriconodonts seem to have lost their reptilian jaw articulation, so the hinge was solely betweeen the squamosal and the dentary as in modern mammals. Although it cannot be proved beyond doubt, we may guess that the articular and quadrate had become the bones of the middle ear, so that the eutriconodonts had a middle ear just like that in modern mammals with a malleus, incus and stapes. In *Trioracodon ferox* there is a small protuberance on the paroccipital process, which may have played a part in supporting the tympanic membrane (Figure 8.4). There is a similar structure in modern

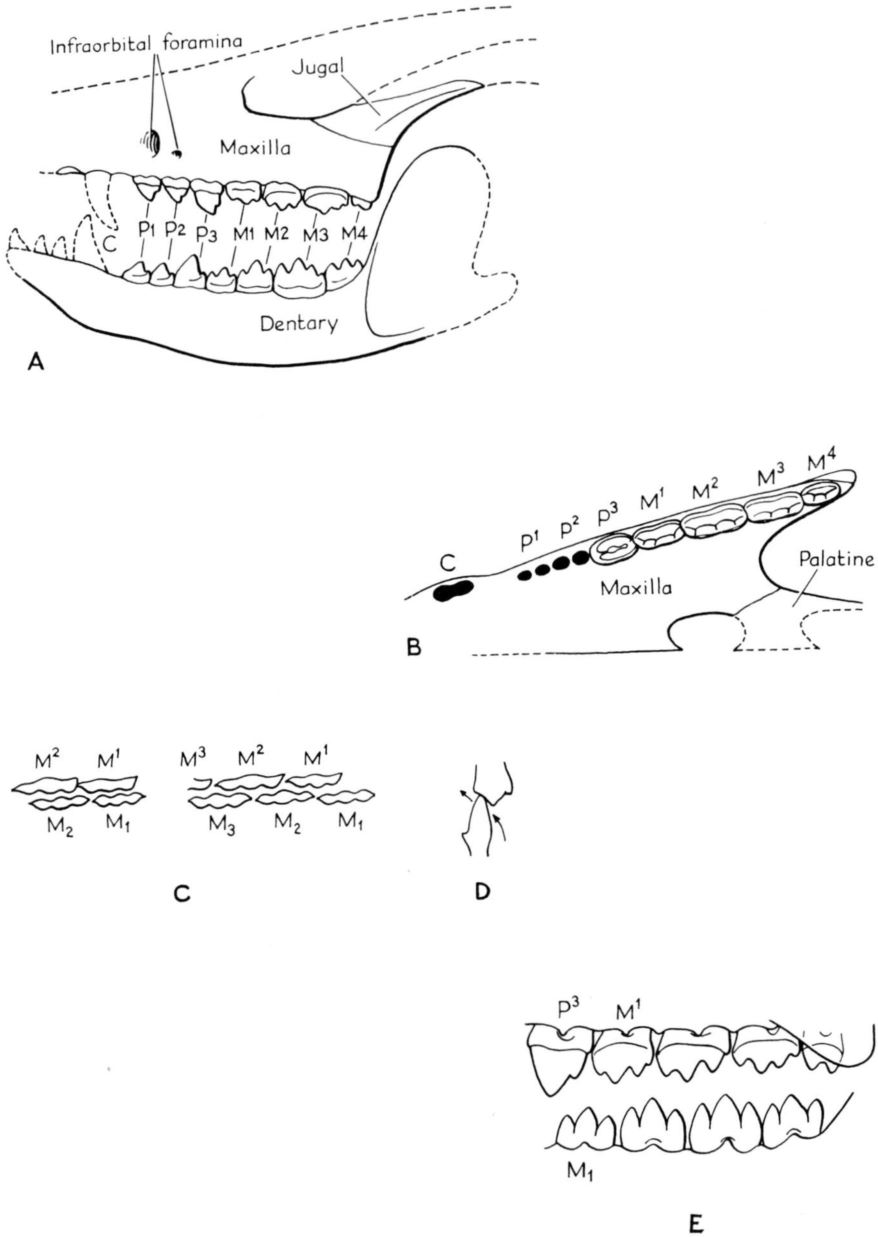


**Figure 8.3:** A, Snout of *Priacodon*; B, Palate of *Priacodon*; C and D, the Occlusion of Triconodont Post-canine Teeth; E, Upper and Lower Post-canine Dentitions
Source: After Simpson, 1925

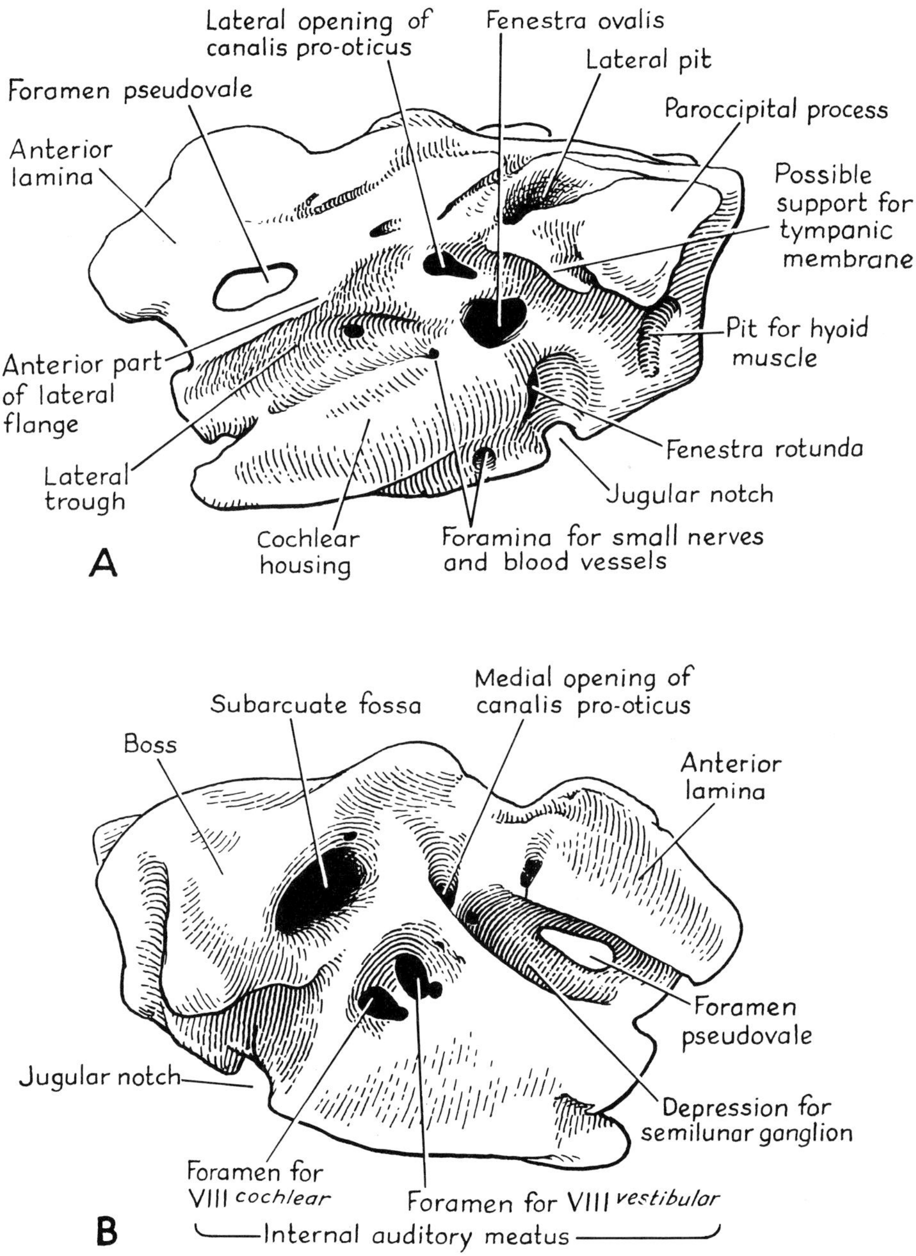


**Figure 8.4:** Petrosal of *Trioracodon ferox*. A, lateral view and B, medial view
Source: After Kermack, 1963

monotremes suggesting that the Eutriconodonta had evolved a fully mammalian type of middle ear (Kermack, 1963)).

The braincase of the eutriconodonts is essentially similar to that of *Morganucodon* in that there is a well-developed anterior lamina (Figures 8.4 and 4.3). The posterior palate also resembles that of *Morganucodon* or the multituberculates in having a well-developed system of longitudinal ridges and troughs (Figure 8.5). In at least some forms, the bony hard palate has a vacuity similar to that found in marsupials but, needless to say, this in no way indicates relationship between these two groups. There is a suggestion that the jugal and lacrimal were reduced, which is also true of the multituberculates but again this cannot be used to indicate relationship: *Morganucodon*, the most primitive member of the Triconodonta known, has a large jugal and a large lacrimal.

The first eutriconodonts were *Amphilestes* and *Phascolotherium* from the Middle Jurassic (Stonesfield Slate) although this is disputed by Mills (1971) who considers them to be aberrant pantotheres (Theria) (Chapter 7). The last known triconodont comes from the Upper Cretaceous (Campanian) of Canada. So the order Triconodonta had an enormous range in time from the first appearance of *Morganucodon* and its allies at the Triassic-Jurassic boundary to the Campanian. This could be as great as 120 $\times$ $10^6$ years making the multituberculates (Table 3.1) their only rivals for longevity among the orders of mammals.

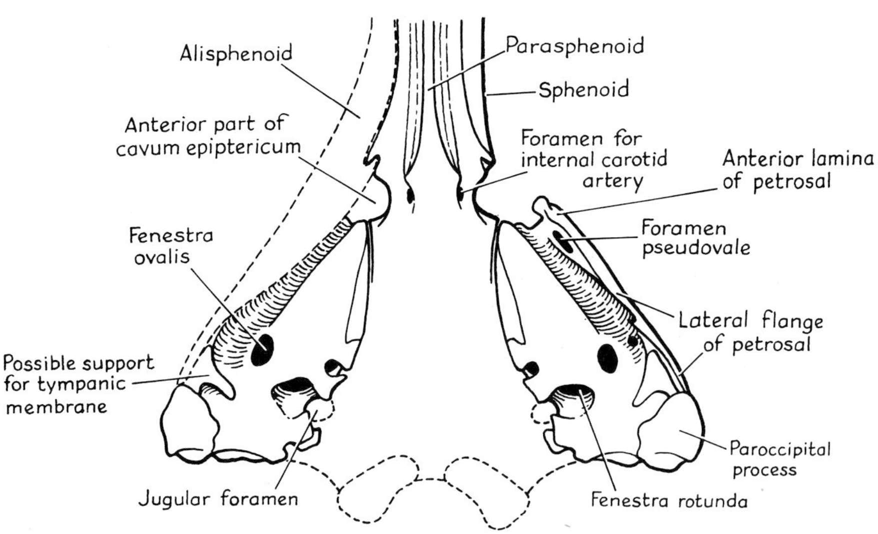


**Figure 8.5:** Reconstruction of a Triconodont Basicranium
Source: After Kermack, 1963

## Docodonta

Turning now to the Docodonta, unfortunately our knowledge of their skulls is less than our knowledge of those of the eutriconodonts, only the jaws of the former being known. Figure 8.6A shows a reconstruction of the lower jaw of *Docodon* from the American Morrison formation. This is almost identical to that of *Morganucodon* and shows no advance upon it. The accessory bones

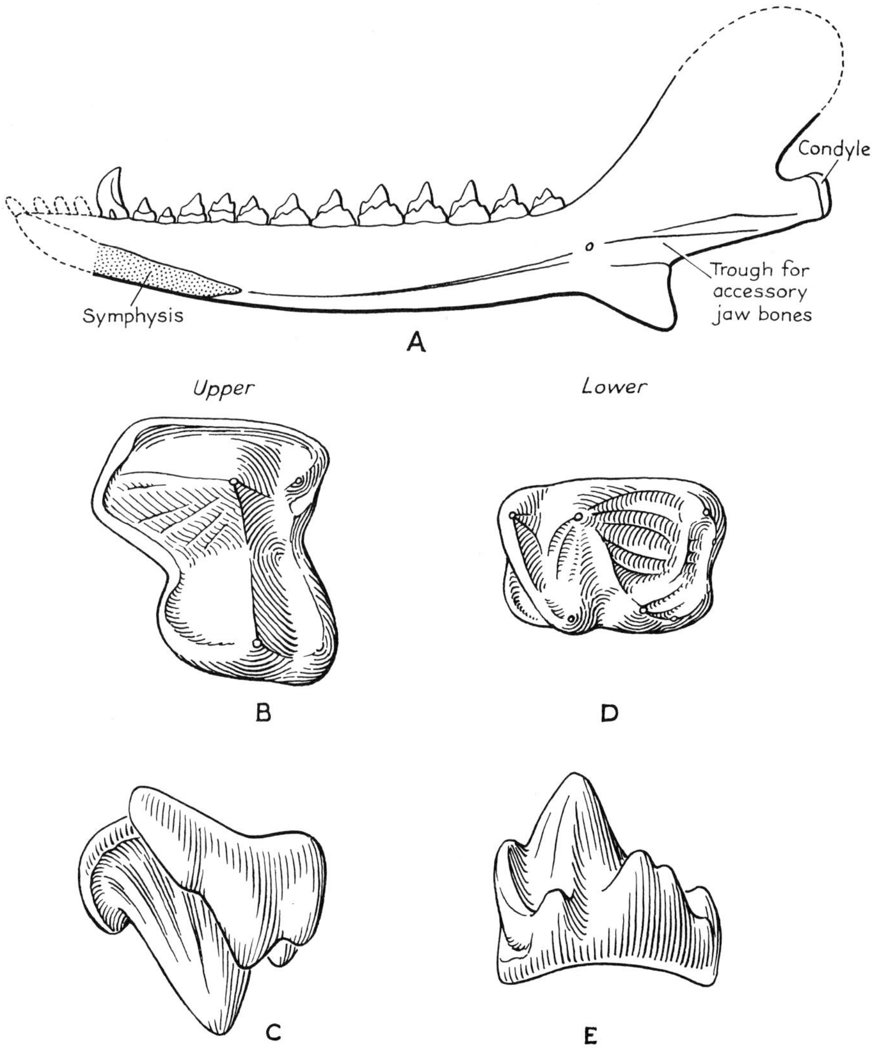


**Figure 8.6:** *Docodon*: A, medial view of the lower jaw; B, and C, occlusal and lingual views of left upper 'molar'; D, and E, occlusal and lingual views of right lower 'molar'
Sources: A: After Simpson, 1929; B, C, D, E: After Vandebroek, 1960

were as large in *Docodon* as they were in *Morganucodon* and there was a double jaw-joint with the quadrate and articular forming a hinge just medial to that formed by the squamosal and dentary. The tympanic membrane or ear-drum was in the lower jaw supported by the reflected lamina of the angular just as it was in *Morganucodon*, or indeed as in the therapsid reptiles. The remarkable thing about the Docodonta is their teeth and these are very remarkable indeed. Recent discoveries and in particular those in Forest Marble (Middle Jurassic) have emphasised just how varied and remarkable docodont teeth were. Simpson (1929) describes docodont molars as 'much the most complicated dental structures known from the Jurassic, or indeed, with the possible exception of some of the Cretaceous multituberculates, from the entire Mesozoic' (Figure 8.6 B to E). His remarks are even more true today than when they were originally made thanks to new discoveries. The Docodonta had reached the same level of functional evolution of their molar teeth in the Jurassic that the Theria attained in the Cretaceous. Unfortunately the recent discoveries have yet to be described, so we shall have to be content with an account of the dentition of *Docodon*. Krusat (1980) gives a full description of *Haldanodon* from the Guimarota.

In the lower jaw of *Docodon* there were three, possibly four, incisors followed by the powerful canine, three or four premolars and seven or eight molariform teeth (Figure 8.6A). In comparison *Haldanodon* from the Guimarota (Kühne, 1968b and Krusat, 1980) has only three lower premolars and five lower molars (Figure 8.7) These premolars are trenchant teeth with prominent backwardly curved main cusps. In *Docodon* the first two upper premolars are similar, but the third is transitional to the molars which lie behind it in the tooth row. This molarisation of the premolariform teeth occurs in advanced placental mammals and is most unusual in the teeth of Jurassic mammals. The molar teeth are extraordinarily advanced in design and parallel in rather a remarkable way the tribosphenic molars which the Theria only attained in the Cretaceous (Figure 8.6 and see Chapter 9).

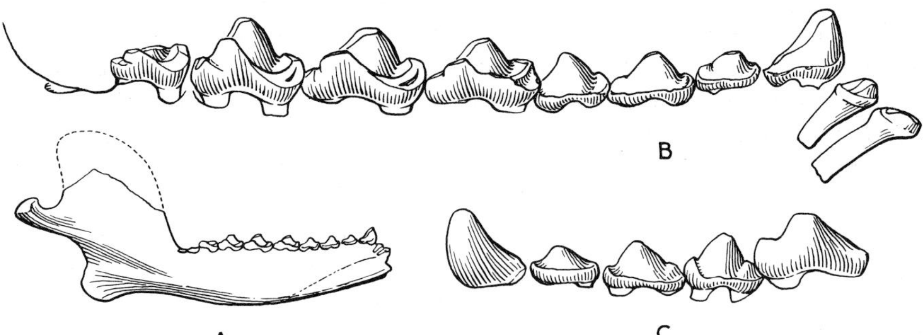


**Figure 8.7:** A, Medial View of Lower Jaw of *Haldanodon*; B, lingual view of dentition (note last post-canine unerupted or lost); C, buccal view of dentition
Source: After Kühne, 1968b

With the increasing knowledge from recently discovered faunas from the Guimarota of the Upper Jurassic, the Forest Marble and Estuarine Series of Skye in the Middle Jurassic, it is clear that the Docodonta were a large component of the mammalian faunas of the Jurassic. This early success they owed to their precocious dentition but their failure to modernise the rest of their anatomy, in particular their possession of the double jaw articulation, proved in the end to be a fatal handicap. It was undoubtedly the retention of the reptilian jaw articulation which prevented their hearing from reaching a fully mammalian level of performance (see Chapter 6). This probably caused the disappearance of the Docodonta soon after the end of the Jurassic as they met increasing competition from the Theria, who were evolving and perfecting the tribosphenic dentition. The very advanced dentition of the Docodonta existing side by side with the primitive jaw articulation is an excellent example of Matthew's Principle (Matthew, 1926; Kermack, 1954). This states that the general level of evolution of related lines evolving in parallel is about the same, and that the line advanced in one character will inevitably be backward in another. The story of the Docodonta shows that a line which becomes so advanced in one character, here the dentition, and yet remains so strikingly backward in another, the jaw articulation and hearing, will in the end be unable to compete with another line which has kept these characters in better balance.

## Multituberculata

The second order, the Multituberculata, first appears in the Upper Jurassic and has not been encountered before, although there is a possibility that there may be representatives in the Middle Jurassic Forest Marble deposits. This suborder of mammals were herbivorous and highly specialised when they are first found in the deposits of Kimmeridgian age in Portugal (Hahn, 1969, 1971, 1978, 1981). Like the Triconodonta they were Atheria, but the two orders must have been separate for a long time. They may well have separated before the acquisition of the squamosal-dentary jaw-joint by their ancestors, in other words, before the latter could be formally described as mammals. The multituberculates may have evolved from the enigmatic haramiyids of the Rhaeto-Liassic (Hahn, 1973), but as the latter are only known by isolated teeth, the comparison is not really particularly useful (Chapter 3).

The Multituberculata hold the distinction of being one of the longest lived mammalian orders. Representatives are found in deposits which range in age from the Upper Jurassic to the Upper Eocene about $110 \times 10^6$ years. Their only rivals are the Triconodonta (Table 3.1). No living order of mammals can equal this, insectivores or marsupials can claim with difficulty no more than $80 \times 10^6$ years. The Mesozoic orders of mammals were in their day highly successful and most of them notably long lived. This contrasts with their usual reputation of being ephemeral and unsuccessful.

The Multituberculata were the first mammals to become herbivorous, replacing the tritylodonts, which had filled the niche of the small chewing herbivore between the top of the Triassic and the end of the Middle Jurassic (some 40 $\times$ $10^6$ years) (Table 3.1). In their turn, the multituberculates continued until they came into competition with the newly evolved rodents early in the Tertiary. This competition with the rodents eventually proved fatal and the multituberculates did not survive to the end of the Eocene. The multituberculate lower jaw consists of a single bone, the dentary with no internal mandibular groove in its medial surface. It will be recalled that an animal with splenial and prearticular bones in the lower jaw needed such a groove running in the dentary to protect blood vessels, a nerve and possibly a forward extension of Meckel's cartilage from being crushed by the accessory bones overlying them. With the loss of the prearticular and splenial, the blood vessels and nerve could quite safely lie along the surface of the dentary.

The interpretation of the multituberculate skull is not easy as, like the present-day monotremes, the sutures of the brain case fuse up early in life. The skull of *Kamptobaatar* (Figure 8.8) from the Upper Cretaceous of Mongolia has been described by Professor Zophia Kielan-Jaworowska (1970, 1971) and is one of the best known multituberculate skulls. As this group of mammals were notably advanced for their time when they first appeared in the fossil record, they did not as a necessary consequence of this, evolve much during their long existence. So we can take *Kamptobaatar* as a model and mention where appropriate how it differs from earlier and later multituberculates.

The dorsal view of the skull (Figure 8.8A) shows that it is rather short for its width with broad powerful zygomatic arches giving it a tritylodont-like appearance. This resemblance is purely functional and does not in any way indicate relationship. Both animals lived the same kind of life; indeed one was the ecological replacement of the other. As in *Morganucodon* and the cynodonts, the zygomatic arches extend to the posterior extremity of the skull. This is also true, but to a much less marked extent in the skulls of living monotremes. Again, this is not indicative of affinity, but the presence of a primitive character that the eutherian mammals have lost.

In side view (Figure 8.8C) the specialisation of the skull of *Kamptobaatar* becomes apparent. The lacrimal and jugal bones are absent; the petrosal has a large anterior lamina like that of *Morganucodon*, whilst the alisphenoid is greatly reduced. No multituberculate has a jugal but a lacrimal is found in *Paulchoffatia* from the Guimarota and this animal is also unusual among the multituberculates in having retained the coronoid accessory bone in the lower jaw.

Figure 8.8B shows the skull of *Kamptobaatar* in ventral view and like the dorsal view it depicts the broad powerful nature of the skull. The palate is broad with no palatal vacuities, although these are found in other multituberculates. The internal nares show ridges formed from the palatines and are similar to those also found in *Morganucodon* and the tritylodonts. The basicranium is broad, almost triangular, in shape, closely resembling that in *Morganucodon*. The petrosal with its anterior lamina lies lateral to the

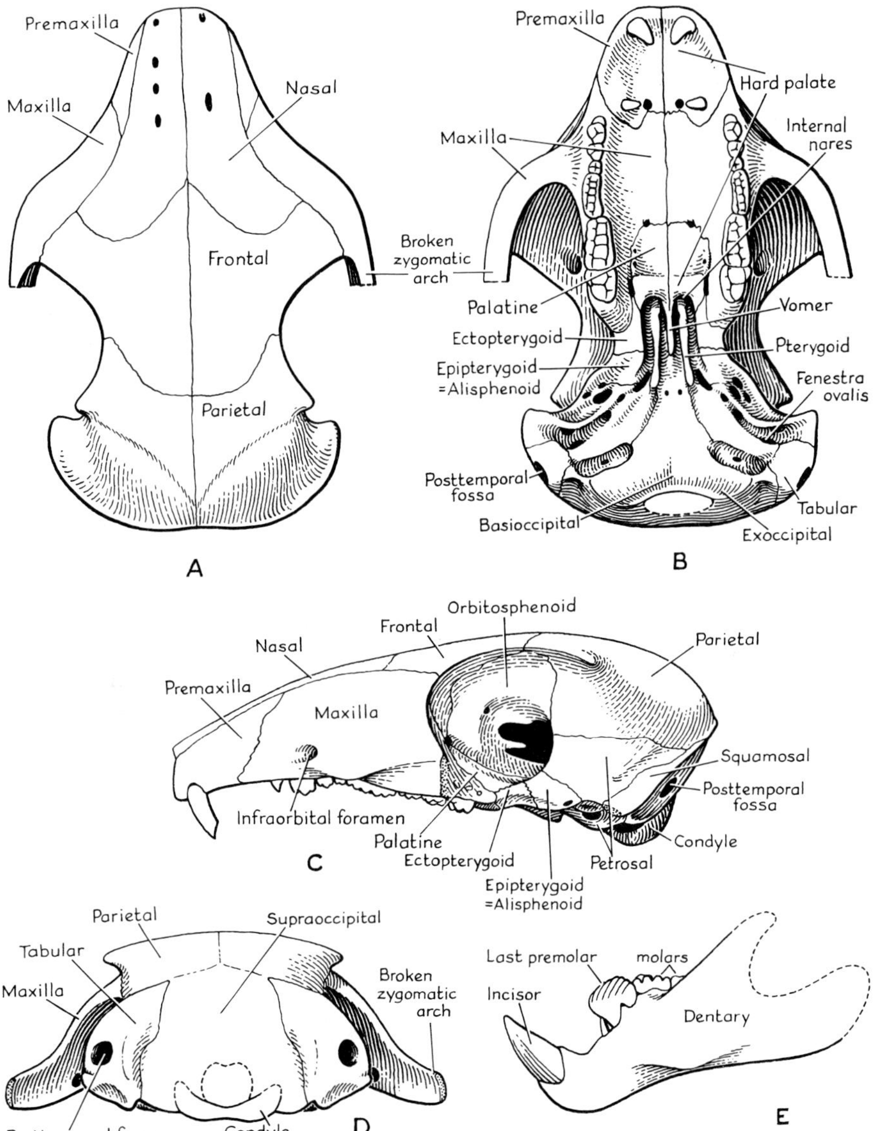


**Figure 8.8:** The Skull of *Kamptobaatar*. A, dorsal view; B, ventral view; C, lateral view; D, occipital view and E, lateral view of lower jaw
Source: After Kielan-Jaworowska, 1971

basicranium, but due to the greater relative width of the skull in multituberculates the lamina is partly on the ventral surface of the braincase, instead of being entirely lateral as in *Morganucodon*. Kielan-Jaworowska has shown that the foramina in the anterior lamina are for branches of the mandibular branch ($V_3$) of the trigeminal nerve. Both the maxillary ($V_2$) and mandibular branches of the trigeminal nerve pass through the anterior lamina in *Morganucodon* and the same may be true for other multituberculates.

The dentition of the multituberculates is highly specialised for dealing with plant food and it becomes more specialised with the passage of time. This is well-illustrated when a series of skulls from the Upper Jurassic through the Cretaceous to the Palaeocene is examined. The series shows the evolution of self-sharpening cropping incisors, the development of a diastema, a reduction in the number of the post-canine teeth and the use of some of the latter for shearing until the development of reasonably efficient cropping incisors. Although the dentition of the Multituberculata is that of a small herbivore, it is fundamentally different from that found in the small mammalian herbivores with which we are most familiar, namely the rodents and lagomorphs, the rabbits and hares. It shows similarities with the dentition of one of the small herbivorous marsupials, in particular that of *Bettongia*, the rat-kangaroo. In the latter, the tritylodonts and multituberculates the incisors are used for cropping vegetable material. But the comparison is not close and it is probably better to regard the dentition of the Multituberculata as unique.

In the earliest multituberculates known, *Paulchoffatia* and *Kuehneodon* from the Guimarota, there are three pairs of upper incisors, with the second pair the largest and the first much the smallest. *Paulchoffatia* is unique amongst the multituberculates in having six upper premolars but *Kuehneodon* had only five, which is the usual number in primitive members of the order (Figure 8.9). The most anterior of these is square in occlusal view with four cusps, more posteriorly the premolars become elongate mesiodistally and there is an increase in the number of cusps. This is most marked in the last premolar which has become somewhat molariform. There are never more than two rows of cusps. In *Paulchoffatia* the premolars are all similar, but the last one, $P^6$, has three rows of cusps and this multituberculate is quite unique in possessing this sixth premolar. There are only two molars in both upper and lower jaws; this is invariable in multituberculates and each tooth has two rows of cusps. The second molar in both genera is positioned markedly lingually with respect to the first. This again is invariable in multituberculates and the reason for it will become clear when the occlusion of the teeth is discussed.

In talking about 'premolars' and 'molars' we are speaking conventionally. Nothing is known of their replacement. Hahn (1978) has suggested that the first 'premolar' is really a canine.

All the Multituberculata have only a single pair of lower incisors and these

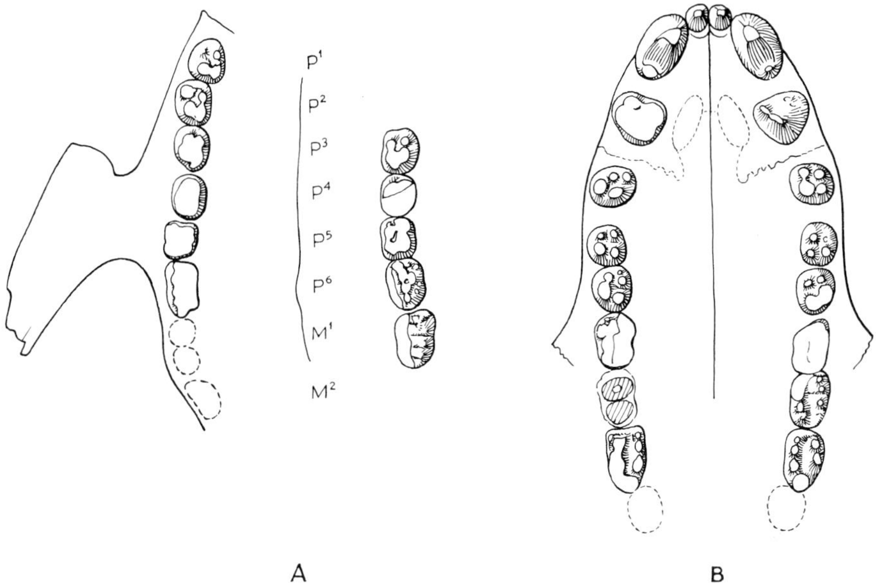


**Figure 8.9:** Upper Dentition of A, *Paulchoffatia delgadoi* and B, *Kuehneodon simpsoni* Source: After Hahn, 1971. A shows dentitions from two different specimens of *Paulchoffatia*

are long procumbent teeth directed forwards. The canines are absent but, in the Jurassic forms, there are four lower premolars and two molars on each side of the jaw (Figures 8.9 and 8.10). No multituberculate has more than this number of lower premolars and molars. The latter are only known from isolated teeth in the Guimarota specimens. The first molar has two rows of cusps, whilst the second is basin-shaped with only a single definitive cusp. This tooth can possibly be matched only in one other multituberculate, a specimen of *Ctenacodon* from the Purbeck.

The lower premolars have a blade-like structure and they shear against the lingual sides of the upper premolars. Hahn considers that in addition to this, there was some chewing action between the upper and lower premolars in the Guimarota multituberculates. Be this as it may, in all succeeding members of the Multituberculata some, at least, of the upper premolars do not come into occlusal relationship with any of the lower teeth. In those upper premolars where occlusion does take place it is limited to the lingual edge of the upper tooth. Figure 8.11 is a reconstruction of the skull of *Ctenacodon* from the Morrison and the Purbeck. This is due to Simpson and gives a good idea of the general appearance of the dentition of a Jurassic multituberculate.

Comparing this dentition with that of the much better known Palaeocene form *Ptilodus* (Figure 8.12) the changes in the dentition are slight but not significant. There are only four premolars instead of the five typical of the Mesozoic multituberculates and the crowns of these teeth are rather broader,

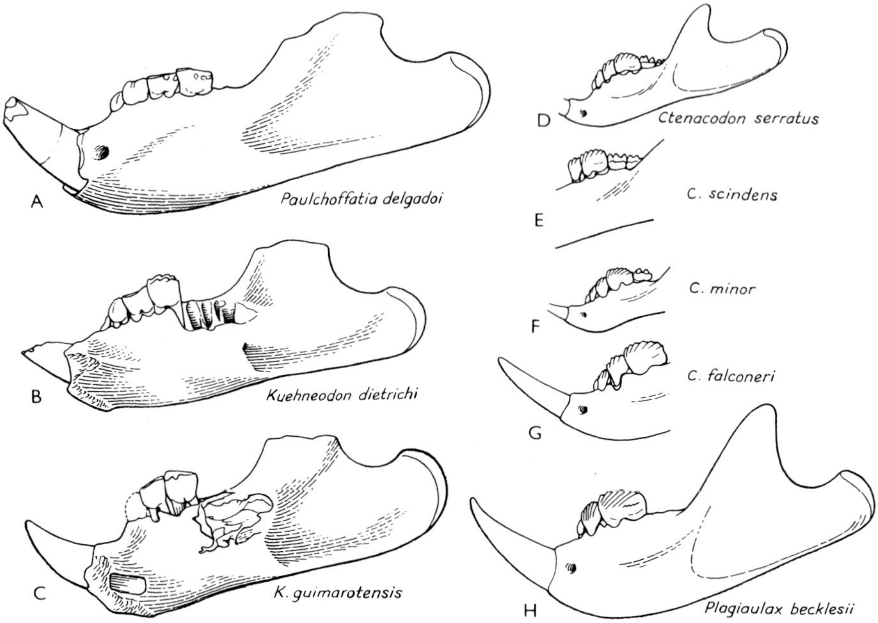


**Figure 8.10:** Lower Jaws of Multituberculates
Source: After Hahn, 1971

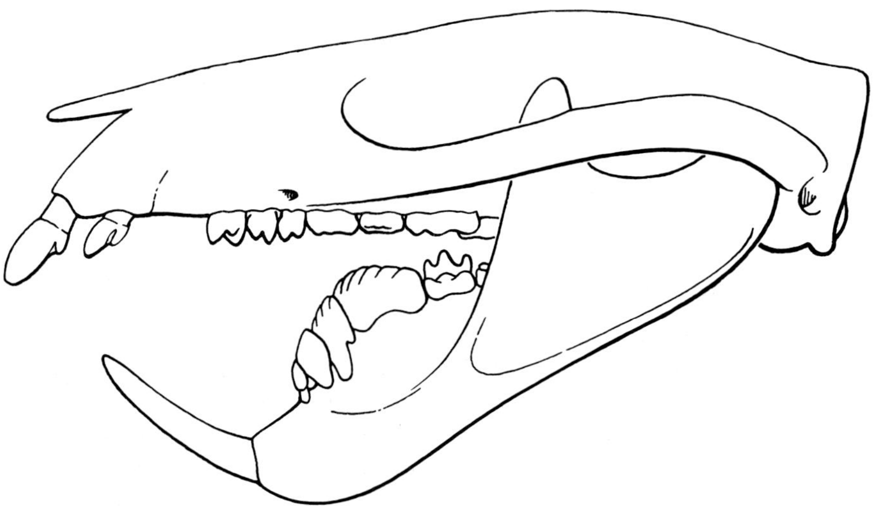

**Figure 8.11:** Reconstruction of the Head of a Primitive Plagiaulacoid – essentially *Ctenacodon*
Source: After Simpson, 1926

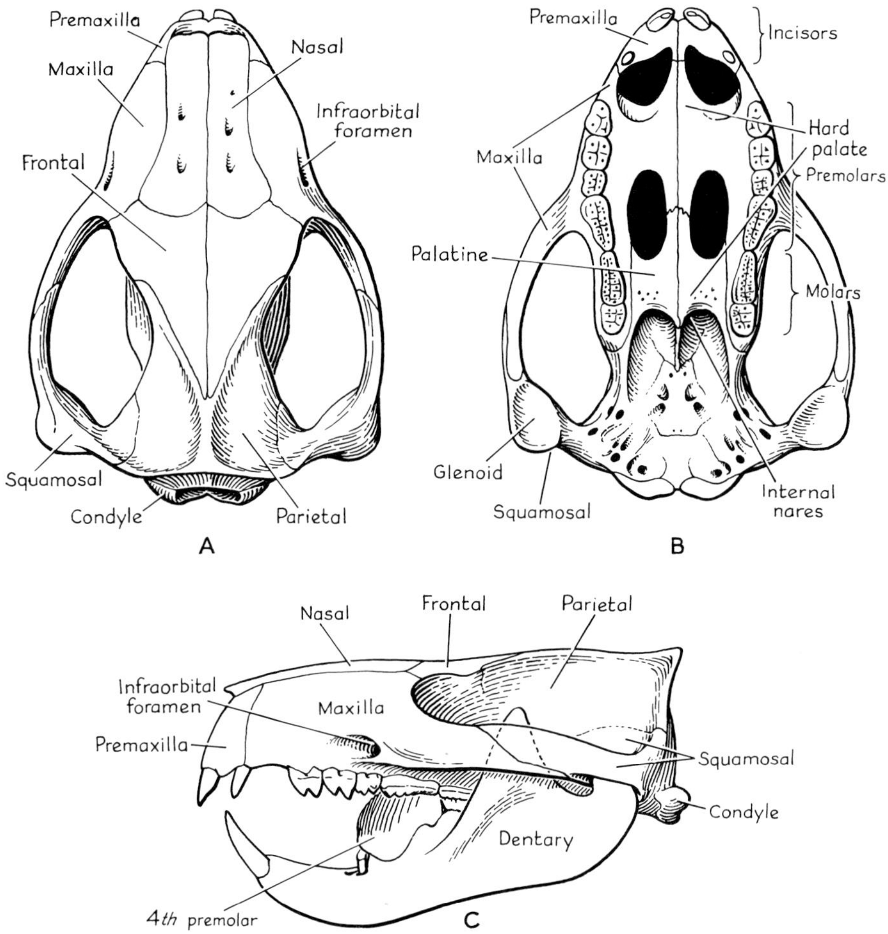


**Figure 8.12:** Reconstruction of the Skull of *Ptilodus*. A, dorsal view; B, ventral view; C, lateral view
Source: After Simpson, 1937

particularly that of the fourth premolar. There are the invariable pair of upper molars on each side, and these have three longitudinal rows of cusps, not two as in the Jurassic representatives of the order. This has been achieved by the addition of a lingual row of cusps to the first molar and a buccal row to the second one. The number of cusps in the first molar has also been greatly increased. In the lower dentition the most obvious change has been the reduction of the premolars to two ($P_3$ and $P_4$) by the loss of $P_1$ and $P_2$. The first of these teeth ($P_3$) is vestigial and reduced to a peg, whilst the second forms an enormous shearing blade. This reduction of the number of lower pre-molars is quite typical of multituberculate evolution. Even in the Upper Jurassic *Plagiaulax* the number of lower premolars has been reduced to three, the first one being rudimentary (Figure 8.10). The lower molars retain the two longitudinal rows of cusps typical of the Mesozoic forms.

The completion of the story is beyond the end of the Mesozoic. In the Palaeocene true rodents first appear, and perhaps under the competition from these rodents, some of the multituberculates evolved a rodent-like dentition themselves. The Lower Palaeocene *Taeniolabis*, for example (Figure 8.13), has evolved the chisel-shaped incisor, growing throughout life from persistent pulps and with the enamel confined to the buccal surface, which is so typical of rodents. Upper and lower incisors work against one another to form a perfect self-sharpening mechanism for gnawing plant material. Much of the success of the rodent is undoubtedly due to the evolution of this dental mechanism. It is worth mentioning that the incisors of typical multituberculates would have been useless for cutting up food into manageable-sized pieces to be further reduced by the molars. This function would have been performed by the shearing premolars. The food was taken into the side of the mouth in a way strangely reminiscent of a carnivore, such as a cat or dog slicing its food up with its carnassial teeth. In *Taeniolabis* the rodent-like incisors would chop up the food to perfection with the consequent loss of the premolars and enlargement of the diastema.*

## Pantotheria

The last and third order of Upper Jurassic mammals are the earliest group of therian mammals, namely the Pantotheria, which diverged into two major sub-groups, the Symmetrodonta and the Eupantotheria. The first known panthothere *Kuehneotherium* occurring in the Welsh fissures has already been described (Chapter 4) and was probably ancestral to the numerous species which radiated out in the Upper Jurassic so that there are more species of Upper Jurassic pantotheres known than there are species of triconodonts or multituberculates (see Table 3.1). The main radiation of the pantotheres has been into two specialised side-branches, the Symmetrodonta and the Dryolestidae, which have absolutely nothing to do with the evolution of the later Theria. Failure to appreciate this has caused much confusion and misunderstanding in the past. Members of these two groups were insectivorous and had dentitions specialised for shearing, using their molar teeth to slice up the hard chitinous cuticles of their prey. The pantotheres were the dominant group of insectivorous mammals in the Upper Jurassic, as the multituberculates were the herbivores and the triconodonts the carnivores. The Docodonta were probably the omnivorous mammals of the day, living on a varied diet, as do modern-day opossums.

*There is a full account of the Mesozoic multituberculates in Clemens and Kielan-Jaworowska (1979).

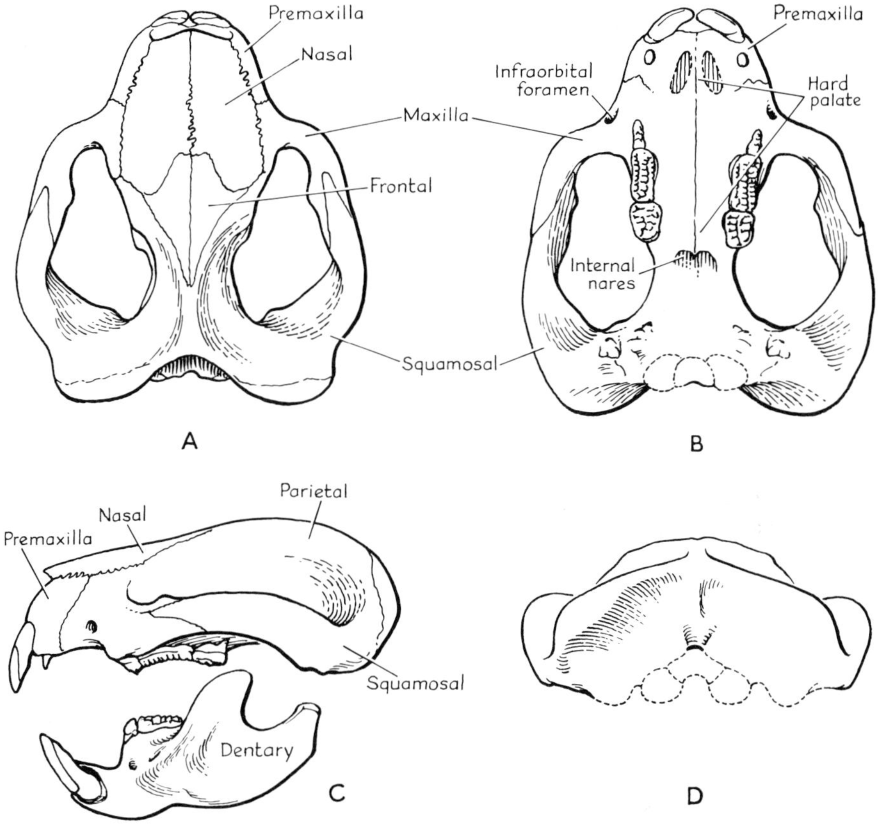


**Figure 8.13:** Reconstruction of the skull of *Taeniolabis*. A, dorsal view; B, ventral view; C, lateral view; D, occipital view
Source: After Simpson, 1937

The Symmetrodonta are the not-greatly altered descendants of *Kuehneotherium*. The changes that have occurred are to increase the efficiency of the molar teeth in shearing by modifications to the cusp arrangements. The symmetrodonts persist into the Middle Cretaceous with little change and although they were unknown before the Upper Jurassic they must have arisen from a form like *Kuehneotherium* in the Lower Jurassic. Since the latter was probably also the ancestor of the eupantotheres, and *Amphitherium*, a typical eupantothere, already exists in the Middle Jurassic (Chapter 7), the symmetrodonts could not have left the main pantothere stock later than the Lower Jurassic. Our failure to find evidence of their existence before the last stages of the Jurassic should be a warning as to how incomplete is our knowledge of mammalian faunas of that period.

In numbers of species the Dryolestidae are the dominant Eupantotheria of the Upper Jurassic, and they could have evolved from a Middle Jurassic form like *Amphitherium*. They have for instance a definite angle for muscle attachment in the lower jaw. Typically they have four premolars and seven molars in their dentitions, which is an increase in number from the seven cheek-teeth occurring in *Morganucodon, Kuehneotherium* and, indeed, in primitive modern mammals. The increase in the number of cheek-teeth shown by the dryolestids is a specialisation, which by itself would preclude them from playing any part in the ancestry of modern mammals. *Amphitherium* also has considerably more than seven post-canine teeth, so whilst it might have been a dryolestid ancestor, it could not have been an ancestor of modern marsupials and placentals.

*Peramus* (Figures 8.14 and 8.15) (Clemens and Mills, 1971) a pantothere from the Upper Jurassic of Swanage, is interesting in that it has only four premolars and four molars with the latter having a larger talonid than any other mammal of certain Jurassic age. These facts suggest it to be the closest-known relative of the later Eutheria (but see *Endotherium*, Chapter 9).

The basic changes needed to turn the molars of *Peramus* into the tribosphenic molars typical of primitive marsupials and placentals (Figure 8.16) are as follows:

(1) the talonid increases in size and becomes basined;
(2) two new cusps appear on the rim of the talonid basin: one, the hypoconid situated buccally; and the other, the entoconid, on the lingual part of the rim;
(3) a new cusp, the protocone, appears in the upper molars and is often the largest cusp in the upper tooth — it lies lingually to the metacone and paracone, which become displaced buccally;
(4) the stylocone becomes reduced in size or disappears;
(5) the protocone comes into occlusion with the cusps forming the rim of the talonid and with the distal face of the metaconid; this produces a 'grinding' action of the teeth, useful in an omnivore.

So in the Upper Jurassic, the early mammals radiated to produce many different species, each of which was capable of exploiting the different types of food material available. Thus there were mammals with shearing teeth for slicing up flesh, others with teeth for cutting through the cuticle of insects and yet others with teeth for masticating plant material to release the nutrient materials entombed in cellulose plant walls. It is unfortunate that more is not known of their skulls and post-cranial skeletons but this lack of material should act as a stimulus for a world-wide search for new material. Although the lack of Upper Jurassic material is to be regretted, we shall see in the next chapter that even less is known of Early Cretaceous mammals. This is unfortunate as the material we have indicates that it is in the Cretaceous that marsupials and placentals appear.

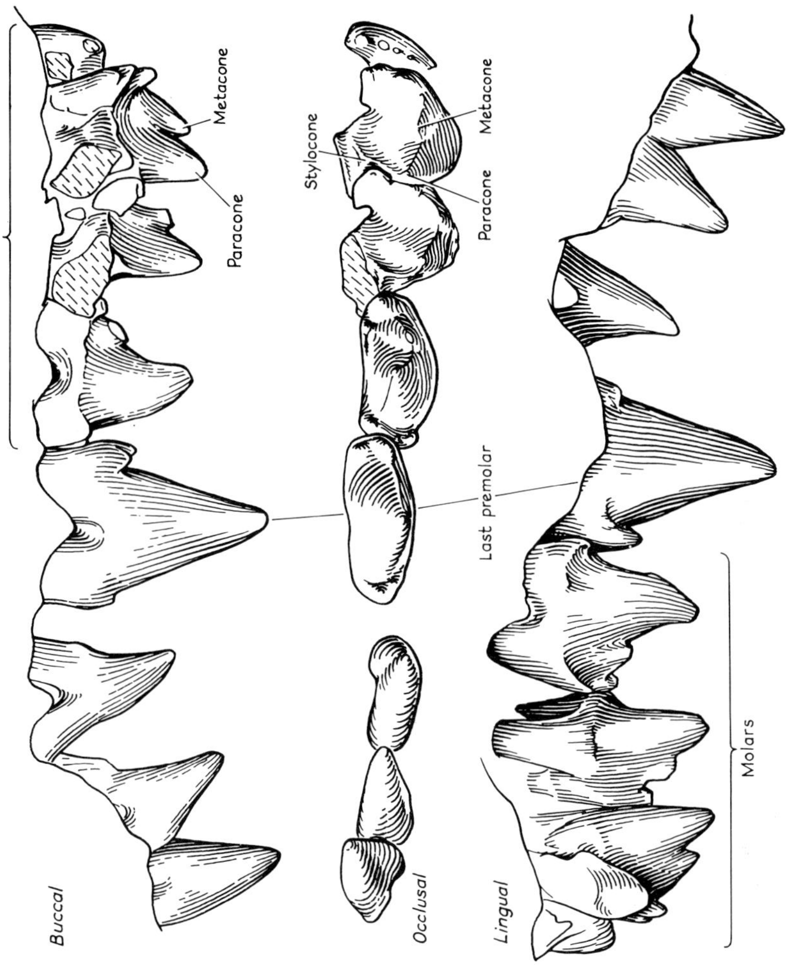


**Figure 8.14:** Upper Dentition of *Peramus*
Source: After Clemens and Mills, 1971

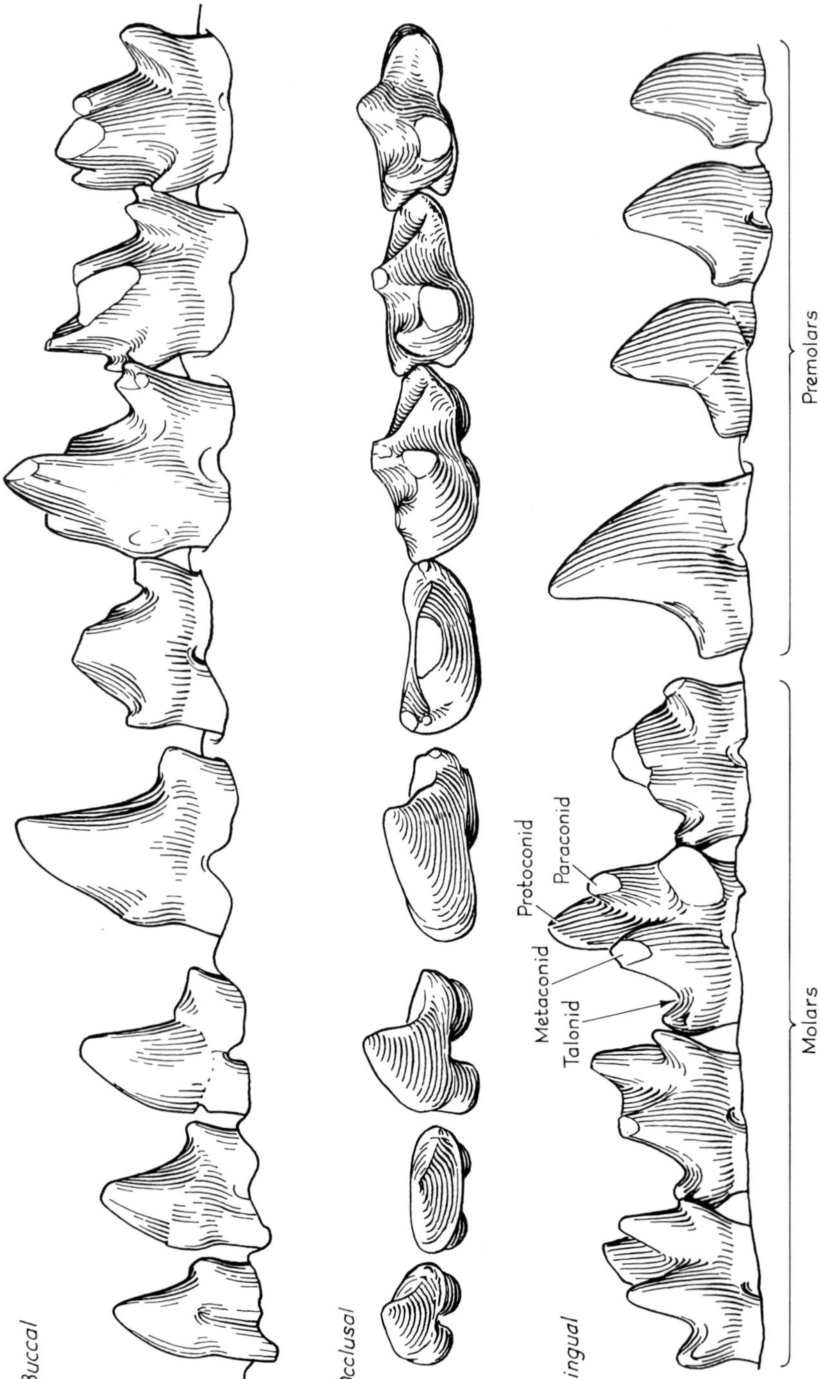


**Figure 8.15:** Lower Dentition of *Peramus*
Source: After Clemens and Mills, 1971

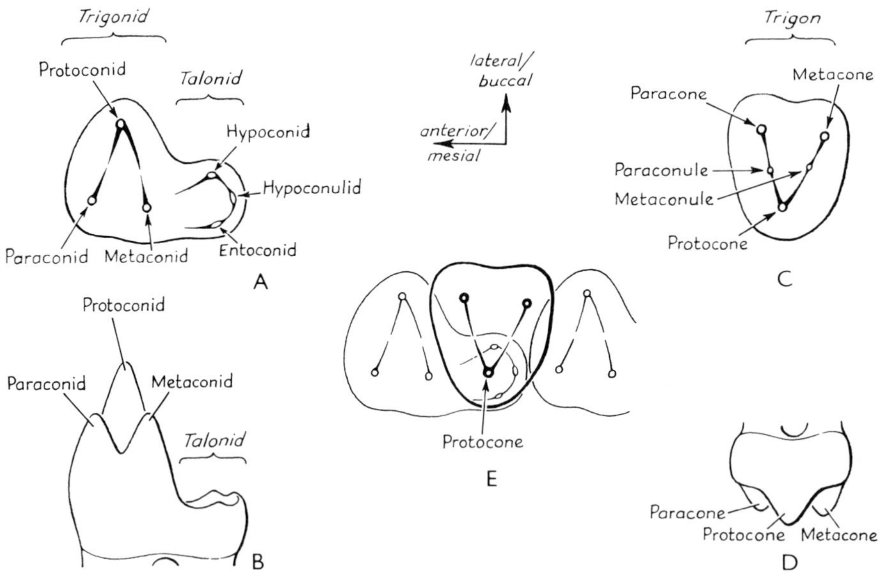


**Figure 8.16:** The Nomenclature of the Cusps of Tribosphenic Teeth. A, lower molar — occlusal view; B, lower molar — lingual view; C, upper molar — occlusal view; D; upper molar — lingual view; E, the teeth in occlusion
Source: After Osborn, 1907

## References

Branca, W. (1916) 'Ein Säuetier? — Unterkiefer aus den Tendaguru-Schichten', *Wissens. Ergeb. d. Tendaguru-Expedition 1909–12. Arch. Biontol., 4*, 137–40

Clemens, W. A. & Kielan-Jaworowska, Z. (1979) 'Multituberculata' in J. A. Lillegraven, Z. Kielan-Jaworowska & W. A. Clemens (eds.) *Mesozoic Mammals*, University of California Press, Berkeley

Clemens, W. A., Lillegraven, J.A., Lindsay, E. H. & Simpson, G. G. (1979) 'Where When and What' in J. A. Lillegraven, Z. Kielan-Jaworowska & W. A. Clemens (eds.) *Mesozoic Mammals*, University of California Press, Berkeley

Clemens, W. A. & Mills, J. R. E. (1971) 'Review of *Peramus tenuirostris* Owen (Eupantotheria, Mammalia)', *Bull. Brit. Mus. (nat. hist) (Geol.), 20*, 87–113

Chow, M. & Rich, T. H. V. (1982) '*Shuotherium dongi genus et species novo*, a Therian with Pseudo-tribosphenic Molars from the Jurassic of Sichuan, China', *Aust. Mamm.*, no. 5, 1–27

Dietrich, W. O. (1927) '*Brancatherulum* n.g., ein Proplacentalia aus dem obersten Jura des Tendaguru in Deutsch-Ostafrika', *Zentbl. Miner. Geol. Paläont.* B, 1927, 423–6

Dietz, R. S. & Holden, J. C. (1970) 'The break-up of Pangea', *Scientific American, 223*(4), 30–41

Hahn, G. (1969) 'Beiträge zur Fauna der Grube Guimarota Nr. 3: die Multituberculata', *Palaeontographica* Abt.A, *133*, 1–100

Hahn, G. (1971) 'The dentition of the Paulchoffatiidae (Multituberculata, Upper Jurassic)', *Separata da Memõria 17 (N.S.) dos Servicos Geológicos de Portugal, Lisboa*

Hahn, G. (1973) 'Neue Zähne von Haramiyiden aus der deutschen Obertrias und ihre Beziehungen zu den Multituberculen', *Palaeontographica* Abt.A, *142*, 1–15

Hahn, G. (1978) 'Die Multituberculata, eine fossile Säugetier-Ordnung', *Sonderbd. naturwiss Ver. Hamburg., 3*, 61–95

Hahn, G. (1981) 'Zum Bau der Schädel-Basis bei den Paulchoffatiidae (Multituberculata; Ober-Jura)', *Senckenbergiana lethaea, 61*, **227–45**

Henkel, S. & Krebs, B. (1977) 'Der erste Fund eines Säugetier-Skelettes aus der Jura-zeit', *Umschau, 77*, 217–18

Kermack, K. A. (1954) 'A Biometrical Study of *Micraster coranguinum* and *M. (Isomicraster) senonensis', Phil. Trans. Roy. Soc.* series B, *237*, 375–428

Kermack, K. A. (1963) 'The Cranial Structure of the Triconodonts', *Phil. Trans. Roy. Soc.* series B, *246*, 83–103

Kielan-Jaworowska, Z. (1970) 'New Upper Cretaceous Multituberculate Genera from Bayn Dzak, Gobi Desert' in Z. Kielan-Jaworowska (ed.) *Results of Polish-Mongol. Palaeont. Expeds., pt. II: Palaeontologica Polonica* no.21, 35–49

Kielan-Jaworowska, Z. (1971) 'Skull Structure and Affinities of the Multituberculata' in Z. Kielan-Jaworowska (ed.) *Results of Polish-Mongol. Palaeont. Expeds., pt. III: Palaeontologica Polonica* no.25, 5–41

Krusat, G. (1972) 'Contribuição para conhecimento da Fauna do Kimeridgiano da Mina de Lignito Guimarota (Leiria, Portugal)', IV Parte: *Haldanodon exspectatus*, Kühne & Krusat (Mammalia, Docodonta), *Memória 27 (N.S.) dos Serviços Geológicos de Portugal, Lisboa*, 1–79

Kühne, W. G. (1961) 'A Mammalian Fauna from the Kimmeridgian of Portugal', *Nature, 192*, 274–5

Kühne, W. G. (1968a) 'Contribuição para a Fauna do Kimeridgiano da Mina de Lignito Guimarato (Leiria, Portugal)', I Parte: History of Discovery, Report on Work Performed, Procedure, Technique and Generalities, *Memória 14 (N.S.) dos Serviços Geológicos de Portugal, Lisboa*, 1–20

Kühne, W. G. (1968b) 'Kimeridge(sic) Mammals and Their Bearing on the Phylogeny of the Mammalia' in Ellen T. Drake (ed.), *Evolution and Environment*, Yale University Press, New Haven

Matthew, W. D. (1926) 'The Evolution of the Horse; a Record and its Interpretation', *Q. Rev. Biol., 1*, 139–85

Mills, J. R. E. (1971) 'The Dentition of *Morganucodon*' in D. M. Kermack and K. A. Kermack (eds.), *Early Mammals, Zool. J. Linn. Soc. Lond., 50*, suppl. 1, 29–63

Osborn, H. F. (1907) *The Evolution of Mammalian Molar Teeth*, Macmillan, London

Owen, R. (1871) *Monograph of the Fossil Mammalia of the Mesozoic Formations*, Palaeontographical Society, London

Simpson, G. G. (1925) 'Mesozoic Mammalia. 1. American Triconodonts: part 2', *Amer. J. Sci.* fifth series, *10*, 334–58

Simpson, G. G. (1926) 'Mesozoic Mammalia. 4. The Multituberculates as Living Animals', *Amer. J. Sci.* fifth series, *11*, 228–50

Simpson, G. G. (1928a) *A Catalogue of the Mesozoic Mammalia in the Geological Department of the British Museum*, British Museum (Natural History), London

Simpson, G. G. (1928b) 'Mesozoic Mammalia XI. *Brancatherulum tendagurense* Dietrich', *Amer. Jour. Sci.* fifth series, *15*, 303–8

Simpson, G. G. (1929) *American Mesozoic Mammalia*, Yale University Press, New Haven

Simpson, G. G. (1937) 'Skull Structure of the Multituberculate', *Bull. Amer. Mus. Nat. Hist., 73*, 727–63

Vandebroek, G. (1960) 'The Comparative Anatomy of the Teeth of Lower and Non-specialized Mammals (volume of plates)', *International Colloquium on the Evolution of Lower and Non-specialized Mammals*, Brussels: Koninklijke Vlaamse Academie voor Wetenschappen, Letteren en Schone Kunsten van België

Willett, E. W. & Willett, H. (1881) 'Notes on a Mammalian Jaw from the Purbeck Beds at Swanage, Dorset', *Q. J. Geol. Soc. Lond., 37*, 376–80

# 9 THE MAMMALS OF THE LOWER CRETACEOUS

The next $40 \times 10^6$ years spanning the Lower Cretaceous was a critical period in mammalian evolution, as it was during this time that the primitive mammals known in the Jurassic evolved into marsupials and placentals. In other words it was in the Lower Cretaceous that modern mammals first appeared; unfortunately much less is known about early Cretaceous mammals than about those from the Jurassic.

## Wealden Mammals

The first Lower Cretaceous mammals were discovered in 1911 at Cliff End, between Hastings and Rye in Sussex, by two Jesuit priests, P. Teilhard de Chardin and F. Pelletier who were at the local seminary. They knew Charles Dawson who gave them the idea of searching for Wealden mammals (Woodward, 1911; and see comments by Dawson when the paper was read at the Geological Society of London). Of the three teeth they collected, one is probably not mammalian but the other two are molars of multituberculates. The teeth were duly described by Smith Woodward, who gave the locality as 'Fairlight'. From the early 1950s we searched for the bone-bed at and near Fairlight Cove without success. In 1960 Professor P. Allen published the correct locality data, which Smith Woodward had disclosed to him in 1938, namely that the bone-bed was at Cliff End about one and half miles along the coast to the east of Fairlight Cove. Later the group from University College, London, collected large quantities of the bone-bed consisting of the hard slabs of limestone packed with teeth and fragments of bone, all cemented together with calcium carbonate. In the laboratory the slabs were broken down in dilute formic acid, which dissolves away the calcium carbonate releasing the bone fragments, the teeth and particles of silica. The residue was washed thoroughly in water and dried. The animal remains were separated from the silica by flotation in a mixture of brominated hydrocarbons, the density of the latter being adjusted so that only the silica floats on the top of the mixture (Lees, 1964). The teeth and bones were then thoroughly cleaned of the brominated hydrocarbons, as these are a serious health hazard. It must be emphasised that this method of extraction and separation of the teeth and bone from the bone-bed can only be undertaken in a suitably well-equipped laboratory under stringent safety regulations and control. The larger fragments of bone are removed by sieving and the final residue consisting almost entirely of fish teeth and bone fragments is sorted by hand under a binocular microscope. To be a successful sorter demands a rare combination of attributes: acute observation allied with the anatomical knowledge to recognise the mammalian teeth, even if they are broken or abraded, has to be combined with the enthusiasm and intellectual drive to keep at the boring and soul-destroying task of examining tens of thousands of unwanted fish teeth to eventually pick out the rare mammalian tooth. On an average one mammalian tooth is found per 200 kilograms of bone-bed. This gives some idea of the magnitude of the task. The majority of Wealden mammal teeth have been discovered by Miss Patricia Lees, who has the necessary skill and determination.

The second Wealden locality was discovered at Paddockhurst Park by Professor J. H. Taylor and a third at Tighe Farm; both these localities have been excavated by the group from University College, London, which holds the major part of the collection of Wealden teeth. A general review of the Wealden mammals has been published by Clemens and Lees (1971).

The fauna from the three localities includes a multituberculate, *Loxaulax valdensis*; a dryolestid, *Melanodon hodsoni* and two symmetrodonts, *Spalacotherium tricuspidens* and *S. taylor*. They are all rather similar to those from the Purbeck and the Morrison (Chapter 8) and it is a little surprising that no triconodonts or docodonts have been found at these localities.

However, there is also one lower molar from Cliff End, which is quite different from anything from the Upper Jurassic. This is *Aegialodon dawsoni* (Figure 9.1) (Kermack *et al.*, 1965). The wear facets on this lower molar indicate that it occluded against a protocone in the corresponding upper molar. If this is so it is the first appearance of this cusp which is so characteristic of the later Theria.

In 1975 Butler and Ford announced the discovery of multituberculate teeth in the Wealden of the Isle of Wight. The teeth are assigned to the genus *Loxaulax* and are of Upper Wealden age — considerably younger than the specimens of the same genus from Cliff End and Tighe Farm.

Within the last two decades Lower Cretaceous mammals have been reported from Spain. The first locality to be discovered was at Galve, in Teruel Province (Crusafont and Adrover, 1966a, b; Kühne, 1966; Crusafont and Gibert, 1976). Mesozoic mammals have also been discovered at Uña, in Cuenca Province, about 60 miles west of Galve (Kühne and Crusafont Pairo, 1968; Henkel and Krebs, 1969; Krebs, 1971). Both localities were worked by breaking up, washing, sieving and sorting large quantities of matrix — the method that has been employed at the majority of Mesozoic mammal localities worked since 1945. It is worth noting that Professor Kühne, to whom, more than to anyone else, the resurgence of interest in Mesozoic mammals in the last thirty-five years is due, was involved in the work at both sites. From both localities have come multituberculates and pantotheres, and they are most likely of Late Wealden age.

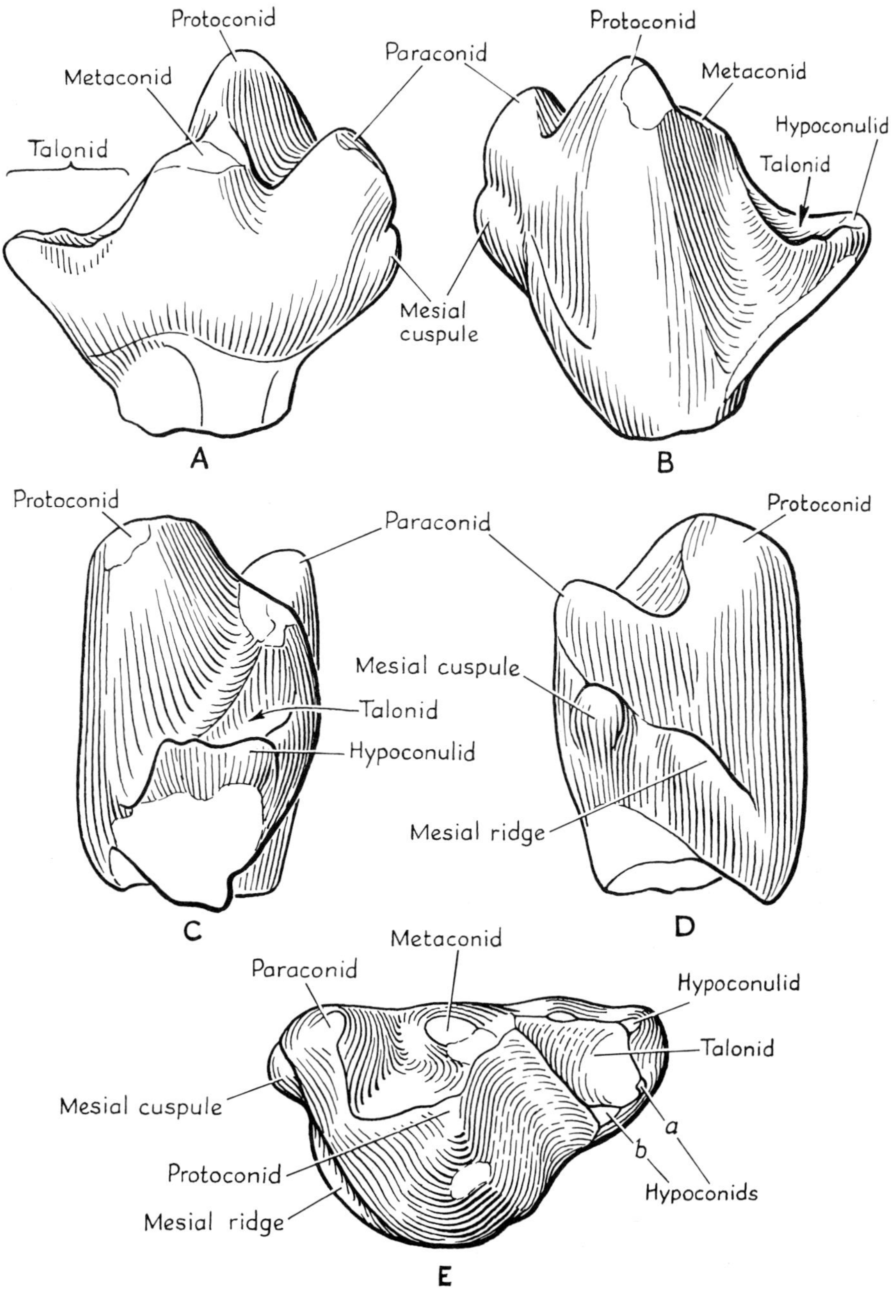


**Figure 9.1:** Lower Molar of *Aegialodon dawsoni*. A, lingual; B, buccal; C, distal; D, mesial; E, occlusal views
Source: After Kermack *et al.*, 1965

## Asian Lower Cretaceous Mammals

Recently mammals have been reported from the Lower Cretaceous of Mongolia. The Aptian-aged Khovboor beds are considerably younger than the Wealden localities and so it is interesting that a tooth called *Kielantherium*, which is *Aegialodon*-like, should be found in them along with triconodonts and multituberculates (Beliajeva *et al.*, 1974; Dashzeveg, 1975; Fox, 1976). This work has yet to be published in full but when it is we should know

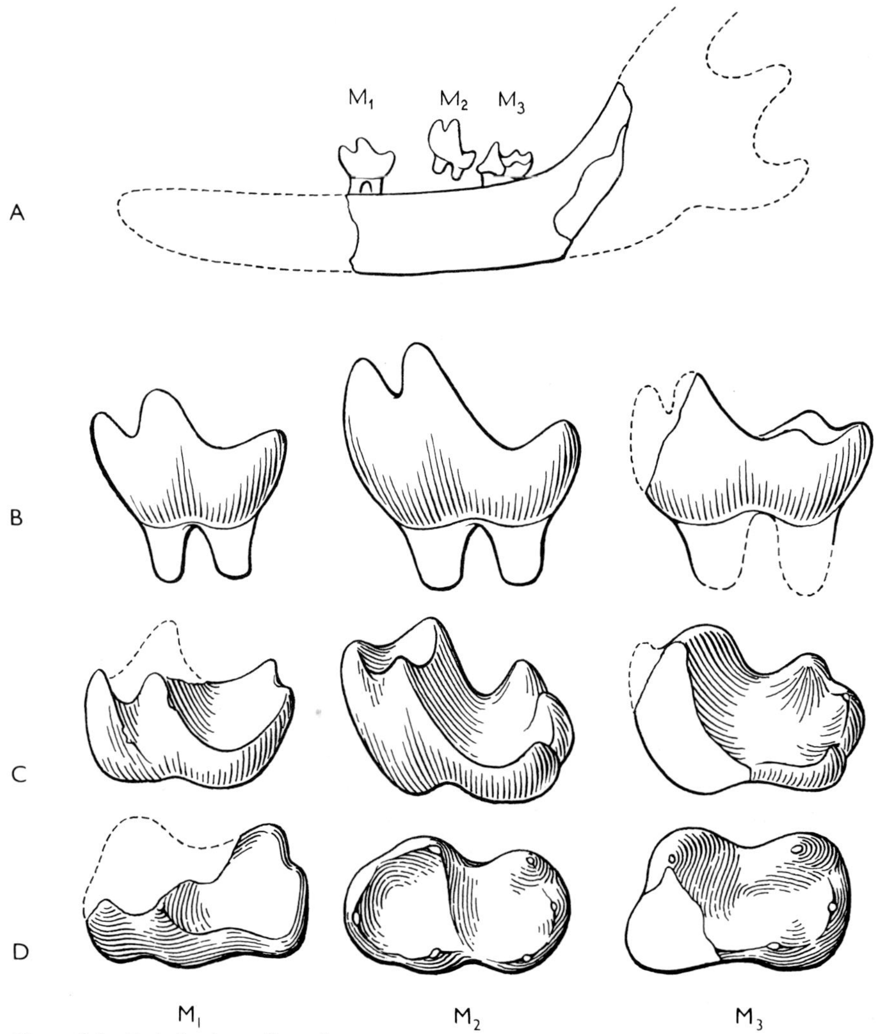


**Figure 9.2:** *Endotherium niinomii*
Source: After Shikama, 1947

more about the origin of therian mammals. Also from East Asia are *Endotherium* (Shikama, 1947) (Figure 9.2) and *Manchurodon* (Yabe and Shikama, 1938), which were found in the Husin coal-field in South Manchuria. The Husin Coal Series is believed to be of Upper Jurassic or possibly Early Cretaceous age, and we think that it is most unlikely to be later than this (Kermack, 1975 — but see Fox, 1972, for a different opinion). The *Manchurodon* tooth is very similar to that of *Amphidon* from the Morrison of USA but *Endotherium* shows dental characters which are otherwise unknown before the Trinity Sands Fauna of Middle Cretaceous age. This is more than 40 $\times$ $10^6$ years later than the Upper Jurassic age ascribed to the coal-field by Yabe and Shikama (1938). The lapse of time is equivalent to that from the Oligocene to the present day (35 $\times$ $10^6$ years). Unfortunately the original specimen has been lost but Shikama's original figure makes it quite clear that *Endotherium* had a fully developed tribosphenic dentition and so was probably 30 $\times$ $10^6$ years ahead of its time (Figure 9.2). Other workers in the field have been most reluctant to accept this but as more discoveries are made it is becoming less and less convincing that the collection of Mesozoic mammals we have from deposits prior to the Upper Cretaceous are a fair example of mammals alive at the time. *Aegialodon* is from the earliest Cretaceous and is far too advanced to have evolved from any known mammal living at the end of the Jurassic in Purbeck or Morrison times.

With our present knowledge and accepting that *Endotherium* is probably of Early Cretaceous or even Upper Jurassic age, as Shikama suggested, then there are two alternative theories to explain the facts. The first is to assume that *Endotherium*-like forms existed in the Early Cretaceous of Europe and North America and that they evolved in these areas. Then *Aegialodon* is the only one of these forms which has been discovered to date. This is a perfectly possible hypothesis. The alternative assumption is that the fully tribosphenic tooth evolved outside the area of Europe and North America. The discovery of *Endotherium* and the *Aegialodon*-like *Kielantherium* in Eastern Asia suggests that this may be the region where the tribosphenic tooth evolved. *Aegialodon* migrated to Western Europe early in the Cretaceous, but the main migration did not occur until the Middle Cretaceous, when a rich fauna of therian mammals with fully evolved tribosphenic teeth is known from the Trinity Sands of Texas. It must be remembered that at this time the Atlantic Ocean was still a narrow strait and migration between Europe and North America was easily possible (see Figures 8.1 and 8.2, Chapter 8).

## The Origin of Marsupials and Placentals

The specimens from the Trinity Sands of Texas are of Albian age and about 100 $\times$ $10^6$ years old, coming from the very top of the Lower Cretaceous. They are therefore younger than the Mongolian teeth which were found in Aptian deposits. The Texan mammals were first described by Professor Bryan Patterson (1956); and the Trinity Sands have subsequently been worked by Professor Bob Slaughter of the Shuler Museum of Palaeontology in Dallas (Slaughter, 1971). Patterson and his co-workers processed 46 tonnes of the Trinity Sands matrix to obtain 300 mammalian specimens, mostly isolated teeth. This is an average of about 150 kg of matrix per tooth and is a perfectly reasonable rate of return for this sort of work. From Cliff End we averaged only one tooth per 200 kg matrix processed and the latter had to be broken down with acid. Patterson's material from Texas could simply be broken down in water and some of the processing was done in a creek on the site. Slaughter has set new standards in persistence, which have yet to be surpassed. At the Butler Farm locality he processed no less than 30 tonnes of matrix to obtain eleven teeth. Slaughter states that more that 200 tonnes have been processed in his search for Trinity mammals. This is a feat which speaks of itself; but only those who have actually carried out these washing and screening techniques truly appreciate its real magnitude.

The Trinity specimens include multituberculates, *Astroconodon* a triconodont, *Spalacotheroides* the youngest known symmetrodont, and therians. The latter are most interesting as they have fully developed tribosphenic teeth, which convinced Slaughter that the separation of the advanced Theria into marsupials and placentals had already taken place before the end of the Lower Cretaceous. He describes the teeth of *Kermackia* (Figure 9.3) which are rather generalised and perhaps the direct descendant of *Aegialodon* of the Wealden along with *Holoclemensia* (Figure 9.4), which he considers to be an early marsupial, and of *Pappotherium* a primitive placental (Figure 9.5).

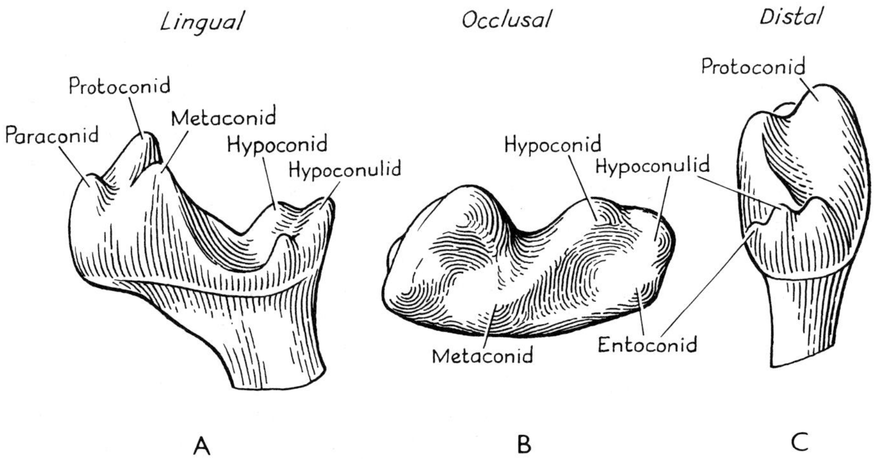


**Figure 9.3:** Lower Molar of *Kermackia*. A, lingual; B, occlusal and C, distal views
Source: After Slaughter, 1971

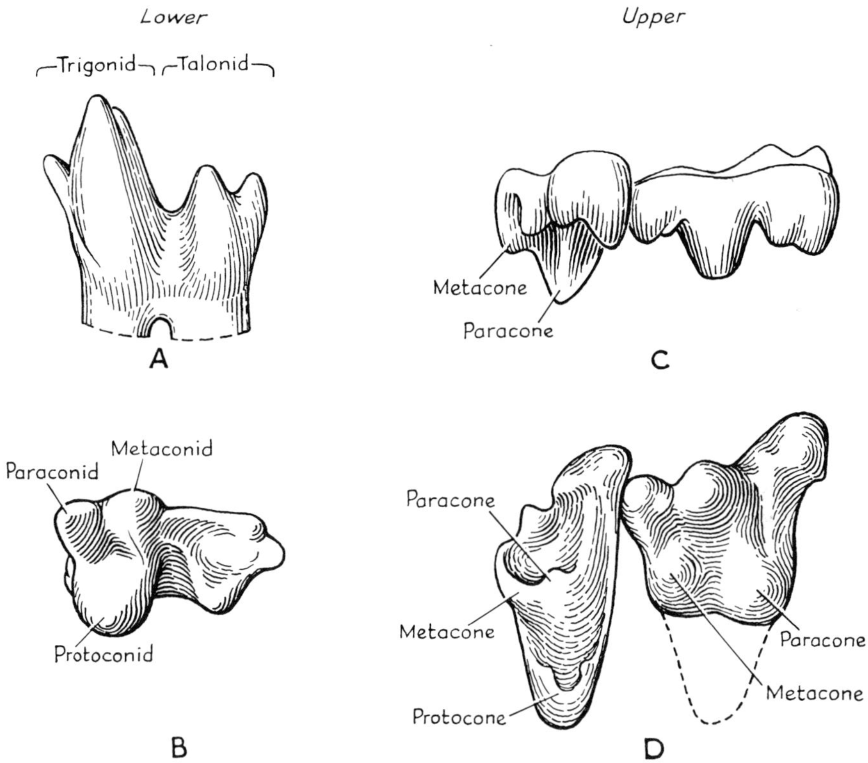


**Figure 9.4:** Molars of *Holoclemensia*. A, buccal; B, occlusal; C, buccal; D, occlusal views
Source: After Slaughter, 1971

Frequently and independently in a number of different lineages, placentals increase the complexity of their premolars until they resemble molars. This 'molarisation of the premolars' is well demonstrated in the modern horse. Marsupials never molarise their premolars. Among his Trinity specimens were two teeth which Slaughter considered to be premolars which had undergone a certain amount of molarisation. Thus he considered them to be the teeth of ancestral placental mammals. He later referred *Pappotherium* to the placentals on the basis of a cusp on the metacrista of the tooth which is similar to that found on the teeth of some insectivores but which is never found in the dentition of marsupials. *Holoclemensia* was classified as a marsupial on the details of the cusp structure of the molar teeth. The structure of these Trinity Sands teeth provided the reasoning for Slaughter believing that ancestral placentals existed in the Lower Cretaceous, so that the placental and marsupial mammals must have separated from one another before that time. Slaughter may well be right, but there are certain assumptions implicit in his argument which should be made explicit. Firstly there is the basic assumption that the placentals and marsupials are the *only* two lineages of higher Theria

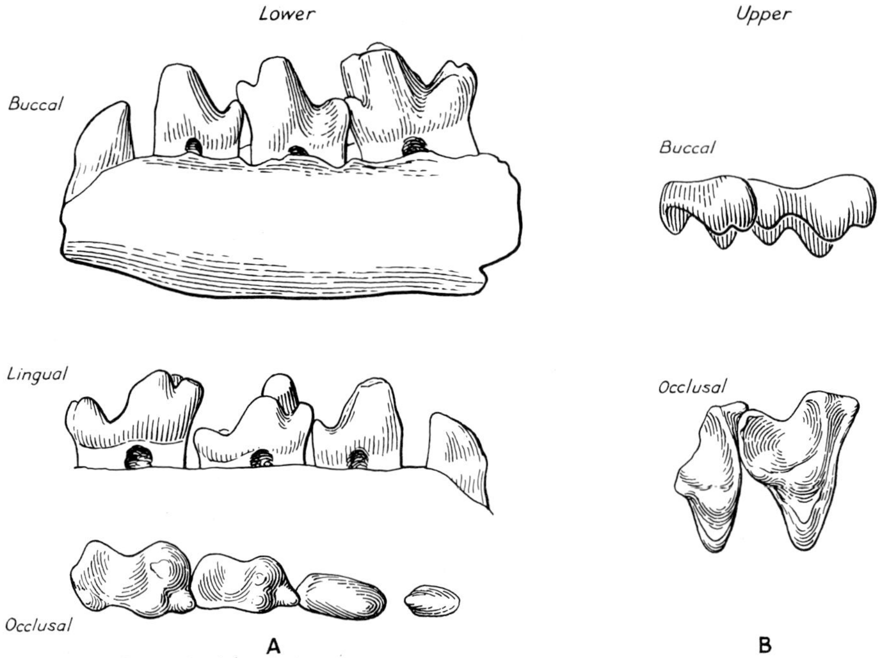


**Figure 9.5:** *Pappotherium*. A, lower postcanine dentition; B, upper molar
Source: After Slaughter, 1971

that have ever existed. Secondly it is assumed that the dental characters which can and are used to define these two lines of mammals in the Tertiary or even in the late Cretaceous can also be used to define them in the late Lower Cretaceous some $30 \times 10^6$ years earlier. Both assumptions are questionable. It is at least as likely that a number of lines of higher Theria arose from the Jurassic pantotheres and that these lines competed with each other so that only two, the placentals and the marsupials, survived on into the later Cretaceous and beyond. At the time that the Trinity Sands were laid down it is possible that Theria existed that were neither marsupials nor placentals but they had dental characters similar to or the same as those which we associate with these two groups of Theria.

## References

Allen, P. (1960) 'Geology of the Central Weald: a Study of the Hastings Beds', *Geol. Ass. Centenary Guide* no.24

Beliajeva, E. I., Trofimov, B. A. & Reshtov, V. (1974) 'General Stages in Evolution of Late Mesozoic and Palaeogene Mammalian Fauna in Central Asia' in N. N. Kramarenko *et al*, (eds.) *Mesozoic and Cenozoic Faunas and Biostratigraphy of Mongolia: Joint Soviet-Mongolian Palaeontological Expedition., Trudy, Palaeontological Institute, Academy of Sciences, Moscow 1*, 19–45 (in Russian)

Butler, P. M. & Ford, R. (1975) 'Discovery of Cretaceous Mammals on the Isle of Wight', *Proc. Isle of Wight Nat. Hist. and Archaeol. Soc., 6*, 662–3

Clemens, W. A. Jr & Lees, P. M. (1971) 'A Review of English Early Cretaceous Mammals' in D. M. Kermack and K. A. Kermack (eds.), *Early Mammals, Zool. J. Linn. Soc. Lond., 50*, suppl. 1, 117–30

Crusafont, M. & Adrover, R. (1966a) 'El primer Mamífero del Mesozoico español', *Publ. Cát. de Paleontología Univ. Barc., 13*, 28–33

Crusafont, M. & Adrover, R. (1966b) 'El primer representante de la clase Maíeros, hallado en el Mesozoico de España', *Teruel, 35*, 139–43

Crusafont, M. & Gibert, J. (1976) 'Los primeros Multituberculados de España (Nota preliminar)', *Acta Geol. Hispan., 11*, 57–64

Dashzeveg, D. (1975) 'New Primitive Therian from the Early Cretaceous of Mongolia', *Nature, 256*, 402–3

Fox, R. C. (1972) 'An Upper Cretaceous Symmetrodont from Alberta, Canada', *Nature, 239*, 170–1

Fox, R. C. (1976) 'Additions to the Mammalian Local Fauna from the Upper Milk River Formation (Upper Cretaceous), Alberta', *Can. J. Earth Sci., 13*, 1105–18

Henkel, S. & Krebs, B. (1969) 'Zwei Säugetier-Unterkiefer aus der Unteren Kreide von Uña (Prov. Cuenca Spanien)', *N. Jb. Geol. Paläont. Mh.*, 1969, 449–63

Kermack, K. A. (1975) 'The Complex of Early Mammals' in *Problèmes Actuels de Paléontologie-évolution des Vertébrés', Colloques Internationaux du Centre National de la Recherche Scientifique, 218*, 563–72

Kermack, K. A., Lees, P. M. & Mussett, F. (1965) '*Aegialodon dawsoni*, a New Tuberculo-sectorial Tooth from the Lower Wealden', *Proc. R. Soc. Lond.*, series B, *162*, 535–54

Krebs, B. (1971) 'Evolution of the Mandible and Lower Dentition in Dryolestids (Pantotheria, Mammalia)', in D. M. Kermack and K. A. Kermack (eds.), *Early Mammals, Zool. J. Linn. Soc. Lond., 50*, suppl. 1, 89–102

Kühne, W. G. (1966) 'Decouverte de Dents de Mammifères dans le Wealden de Galve', *Teruel, 35*, 159–61

Kühne, W. G. & Crusafont Pairo, M. (1968) 'Maíeros del Wealdiense de Uña, cerca de Cuenca', *Acta Geol. Hispan., 3*, 133–4

Lees, P. M. (1964) 'The Flotation Method of Obtaining Mammal Teeth from Mesozoic Bone-beds', *Curator, 7*, 300–6

Lillegraven, J. A., Kielan-Jaworowska, Z. & Clemens, W. A. (eds.) (1979) *Mesozoic Mammals*, University of California Press, Berkeley

Patterson, B. (1956) 'Early Cretaceous Mammals', *Fieldiana, Geol., 13*,(1), 1–100

Shikama, T. (1947) '*Teilhardosaurus* and *Endotherium*, New Jurassic Reptilia and Mammalia from the Husin Coal-field, South Manchuria', *Proc. Japan Acad., 23*, 76–84

Slaughter, B. H. (1971) 'Mid-Cretaceous (Albian) Therians of the Butler Farm Local Fauna, Texas' in D. M. Kermack and K. A. Kermack (eds.), *Early Mammals, Zool. J. Linn. Soc. Lond., 50*, suppl. 1, 131–43

Woodward, A. S. (1911) 'On Some Mammalian Teeth from the Wealden of Hastings', *Q. J. Geol. Soc. Lond., 67*, 278–81

Yabe, H. & Shikama, T. (1938) 'A New Jurassic Mammalia from South Manchuria', *Proc. Imp. Acad. Japan, 14*, 353–7

# 10 EPILOGUE

In the Upper Cretaceous unequivocal marsupials and placentals were in existence, and the early stages of mammalian evolution, which form the subject of this book, are definitely over. Hence only a brief mention of the faunas of the Upper Cretaceous will be made here. For those interested in these faunas, there is a good account and full references to the literature in *Mesozoic Mammals* (1979) (Lillegraven, Kielan-Jaworowska and Clemens (eds.), University of California Press, Berkeley).

The Upper Cretaceous faunas are much richer than those of the Lower Cretaceous, and they give us much more information about the mammals they contain than do those of the earlier periods with which this book is concerned. The main faunas come from Mongolia and North America. The former are rich in multituberculates and placentals and include only one possible marsupial whose identification is in considerable doubt. The North American faunas were quite different. The collections contain abundant multituberculates and marsupials but the placentals are not conspicuous except in the Bug Creek fauna, which is of Uppermost Cretaceous age. There is one mammal, a placental, known from the Upper Cretaceous of Europe and a small marsupial fauna is known from South America. The last of the triconodonts, *Alticonodon lindoei*, is found in the Upper Cretaceous Milk River deposits of North America.

Now in the Upper Cretaceous the Theria can be classified unequivocally as either placentals or marsupials and so the early period of the history of the mammals is completed. The marsupials are almost entirely restricted to North America whilst the placentals are also found in Asia.

With the end of the Cretaceous ($65 \times 10^6$ years ago) we reach perhaps the most spectacular of the great extinctions. It occurred much more rapidly than the great extinctions of terrestrial vertebrates in the Triassic period, and affected marine as well as terrestrial animals. The papers on this Cretaceous extinction are most numerous as are the hypotheses they propound, but none of these theories compels belief. We shall accept the fact of the extinction without speculation on its causes.

Among the groups which became extinct were the two orders of dinosaurs — the Saurischia and the Ornithischia: so much so that the presence of dinosaurs in a deposit is taken as an infallible criterion that the beds are of Mesozoic and not of Tertiary age. Although the dinosaurs disappear completely the birds survive and flourish. This is surprising since the birds are clearly closely related in some way to the two dinosaurian orders; and were the birds a totally extinct group they would be classified, along with the Saurischia and the Ornithischia, as an order of archosaurian reptiles.

The mammals were much less adversely affected than the dinosaurs by the crisis at the end of the Cretaceous. The most significant change was the extinction of many genera of marsupials in North America, so that never again have marsupials a major role in the mammalian faunas of North America.

The mammals of the Palaeocene were mostly of small size. In the late Palaeocene some are larger, the largest being about the size of an ox. The birds, which had survived unscathed where their close relatives the dinosaurs died out, radiated in the Palaeocene to produce giant flightless carnivorous forms. These were the largest carnivores of their time, and must have preyed on the Palaeocene mammals. The evolution of more efficient placental carnivores in the Northern Hemisphere during the Eocene ended the reign of these giant carnivorous birds. In South America, where there were originally no placental carnivores, the giant carnivorous birds survived until the fissipedes entered the continent from North America in the Pliocene.

Turning to the known distribution of mammals in Palaeocene times, we find that in North America the placentals are the dominant group, although some marsupials lingered on and actually succeeded in reaching Europe in the Eocene. Primitive placentals had moved south and reached South America by the Palaeocene and radiated out to form the dominant herbivores of the time. The primitive carnivores following behind the herbivores did not reach South America before the link with North America was broken at the end of the Cretaceous (Figure 8.2); so the marsupials and giant birds radiated out to fill the vertebrate carnivore niche. As yet Palaeocene faunas are unknown in Antarctica and Australia.

From our knowledge of the distribution of placental faunas the story of their evolution seems reasonably clear. The centre of radiation was in the Old World, most likely in Eastern Asia, and their main invasion of the New World was not until the end of the Cretaceous, when they became the dominant mammals of North America and the dominant herbivores of South America. As Australia had separated from Gondwanaland before the placentals reached it, the marsupial fauna has remained dominant there to the present day. Fossil-hunting in Antarctica today is thwart with difficulties, so it is not known if the placentals reached this continent before it separated from South America. It is quite possible that they did. It is also unknown if the placentals had been able to colonise Africa and Peninsula India by this route before these continents detached themselves from Gondwanaland.

The centre of radiation of the marsupials is less obviously located, indeed there are several equally plausible hypotheses. The first and oldest is that the marsupials arose earlier than the placentals and as the latter spread from their centre of radiation in Asia, they exterminated the marsupials and replaced them ecologically. Support for this hypothesis comes from the observation that, at the present time, marsupials are usually unable to compete with placentals. However, there is no palaeontological evidence to support this hypothesis, and against it is the fact that marsupials are unknown in the Old World until the Eocene, when they have quite clearly invaded Europe from North America. This hypothesis is now generally rejected. The second hypothesis is that marsupials arose in North America, colonised South America and from there passed through Antarctica to Australia. There is a third hypothesis, which has attractions. This is that the marsupials had their centre of radiation somewhere in Gondwanaland and colonised North America through South America. The consensus would be that the truth lies between the second and third hypotheses; although a great deal more field-work is required before the problem of the place of origin of the marsupials can finally be solved.

With the great radiation of the placentals at the end of the Cretaceous our early history of the mammals has come to an end. The wheel has come full circle and once again a theropsid group, the placentals, are the dominant group of large land animals, just as were their ancestors, the mammal-like reptiles in the Permo-Trias. The long period of humble obscurity is over and the mammals have at long last come into their own. It is important to remember, however, that the early history of the mammals, from their beginning at the end of the Triassic till the end of the Cretaceous, covers two-thirds of their total history and spans over $140 \times 10^6$ years. This is twice the time-span from the end of the Cretaceous to the present day.

# INDEX